Magnetic
FEVER

*Global Imperialism and Empiricism
in the Nineteenth Century*

TRANSACTIONS

of the

AMERICAN PHILOSOPHICAL SOCIETY

Held at Philadelphia

For Promoting Useful Knowledge

Volume 99, Part 4

Magnetic
FEVER

*Global Imperialism and Empiricism
in the Nineteenth Century*

CHRISTOPHER CARTER

AMERICAN PHILOSOPHICAL SOCIETY
Philadelphia • 2009

ISBN: 978-1-60618-994-8

US ISSN: 0065-9746

Cover illustration:
Toronto Magnetic Observatory (1852) by William Armstrong
© Her Majesty the Queen in Right of Canada, Environment Canada [2007]
Photo by: Olena Koursa
Reproduced with the permission of the Minister of Public Works and Government Services Canada.

Library of Congress Cataloging-in-Publication Data

Carter, Christopher Ray, 1974-
  Magnetic fever : global imperialism and empiricism in the nineteenth century / Christopher Carter.
    p. cm. -- (Transactions of the American Philosophical Society held at Philadelphia for promoting useful knowledge; v. 99, pt. 4)
  Includes bibliographical references and index.
  ISBN 978-1-60618-994-8
  1. Geophysics--Research--Great Britain--History--19th century. 2. Geophysics--Research--United States--History--19th century. 3. Science and state--Great Britain--History--19th century. 4. Science and state--United States--History--19th century. 5. Research--International cooperation--Great Britain--History--19th century. 6. Research--International cooperation--United States--History--19th century.
  I. Title.
  QC807.6.G7C37 2009
  550.72'041--dc22

                                                              2009025414

Dedicated to my mother,
Betty H. Carter

Without whom none of it
would have been possible

And in memory of my father,
Bobby R. Carter

John Herschel

Table of Contents

Preface

In the nineteenth century, geophysics and empire building developed in tandem. As a result, empirical science could utilize the imperial structure to conduct research on a worldwide scale. Scientific motives for such studies included the need to confront both weaknesses in the inductive approach to science as well as the reality of new scientific fields that could only be studied globally. In Britain and the United States, the interaction of science and state allowed a range of geophysical projects to develop. In both countries, scientists had to find ways to overcome the difficulties that made their governments hesitant to support such ventures. In Britain, John Herschel and Edward Sabine provided a crucial connection between the philosophical, political and social elements that made this research possible. At Herschel's instigation, colonial observatories were added to the British venture for an Antarctic expedition known as the Magnetic Crusade, initiating a worldwide system of physical observatories conducting a coordinated series of continuous observations. The British system expanded to include new outposts in America as well as throughout the British domains. As a result, new sciences such as geomagnetism and meteorology developed more fully than they could have when only local observations were available. I argue that the empire provided a setting where universal science could be practiced and legitimized, helping both to overcome the inherited problems of the inductive method and setting up a system by which scientists could study interconnected phenomena on a global scale.

Acknowledgments

There are many people who assisted me at various stages in the production of this book and to whom I am indebted. First, I would like to thank my graduate advisor Seymour Mauskopf, for suggesting the original idea for this study and reading through countless drafts of chapters. I also thank the other members of my dissertation committee: Alex Roland, Cynthia Herrup (who suggested the title) and Michael McVaugh for their input and comments.

As the son of an archivist, I must acknowledge the people who helped me the most in finding the sources that I needed. The staff at the National Archives (formerly the Public Record Office) in Kew and the Manuscript Department at St. Andrews University Library were invaluable. Specifically, I must express my gratitude to Tara Wegner at the Harry Ransom Center in Austin, Solveig Berg at the Institutt for Teoretisk Astrofysikk in Oslo and Marc Rothenberg at the Smithsonian Institution for all of their help. Additionally, I am grateful to John Herschel-Shorland for kindly granting permission to copy materials from the Herschel archives.

Finally, I am thankful for the assistance of Mary McDonald of the American Philosophical Society and the anonymous reviewers of my manuscript for working with me on revisions of the text over the last two years.

List of Abbreviations

APS   American Philosophical Society, Philadelphia

BL   British Library
       Add.—Additional Manuscripts

DAB   Dictionary of American Biography

DNB   Dictionary of National Biography

DSB   Dictionary of Scientific Biography

DU   Duke University

FC   Faraday Correspondence (James, ed.)

ITA   Institute for Theoretical Astrophysics, Oslo

JHP   Joseph Henry Papers (Rothenberg, ed.)

NLS   National Library of Scotland, Edinburgh

PRO   Public Record Office, London
       (now National Archives)

       BJ1 – Kew Observatory Papers

       BJ2 – Ross Papers

       BJ3 – Sabine Papers

       BJ7 – Fitzroy Papers

       BJ9 – Meteorological Department Papers

       RGO – Royal Greenwich Observatory Papers

RS   Royal Society, London

       HS – Herschel Papers

       MM – Miscellaneous Manuscripts

       Sa – Sabine Papers

       Te – Terrestrial Magnetism Archive

SAUL   St. Andrews University Library, St. Andrews

SIA   Smithsonian Institution Archives, Washington DC

UTX   University of Texas, Austin

WC   Whewell Correspondence (Todhunter, ed.)

Introduction

On September 25, 1841 instruments from Greenwich to St. Helena and from Toronto to Munich recorded a massive geomagnetic disturbance. Compass needles swung from north to east and brilliant aurora brightened the early morning skies. At Greenwich, the needle moved more than two and a quarter degrees in less than eight minutes.[1] German observers confirmed the "great magnetic perturbation."[2] To these spectators, it was as if the planet's magnetic field had been temporarily thrown out of balance. By December, there was enough information to conclude that the "storm" appeared to "have taken place simultaneously over the whole surface of the globe."[3] No one could explain its cause or find any comparable record of such a violent magnetic occurrence. George Airy, the Astronomer Royal at Greenwich, later commented that he could not "conceive what was happening to terrestrial magnetism on September 25 to make disturbances everywhere so violent yet so discordant."[4]

In 1841, the unprecedented event was not the worldwide magnetic disturbance, but the fact that it was noticed at all. Before computers and satellites, such magnetic disruptions could only be detected through the observation of magnetized needles at geophysical observatories. Only two years before, the British government had sponsored a scientific expedition known to history as the Magnetic Crusade.[5] In

---

1. Memo by George Airy, October 26, 1841. (PRO BJ3/20/10).
2. Lamont to Sabine, December 15, 1841. (PRO BJ3/4/43); Gauss to Sabine, Mach 14, 1842. (PRO BJ3/2/78).
3. Sabine to H. D. Ross, December 24, 1841. (PRO BJ3/79/39–42).
4. Airy to Sabine, January 10, 1842. (PRO BJ3/20).
5. Although several historians have suggested that the term "Magnetic Crusade" was contemporary, it is unclear if it was ever used by the major figures lobbying for the project. Jack Morrell and Arnold Thackray, *Gentlemen of Science* (Oxford, Clarendon Press 1981), 517; John Cawood,

addition to voyaging to the Antarctic in search of the southern magnetic pole, the crusade also established a number of physical observatories in British colonies from Canada to the Cape of Good Hope. The
system of observations they created tied into and expanded existing
observations being made in Europe. By the 1840s, this international
network of observatories, later called "the greatest scientific undertaken which the world has ever seen," had been established.[6] This network recorded the great magnetic storm of 1841. The significance was
that these observatories were spread out over thousands of miles, and
yet recorded the same storm. For the first time, scientists realized that
magnetic storms could affect the entire world simultaneously and that
they could now study such phenomena on a global scale.

Studying geomagnetism, meteorology, or the tides in the nineteenth century was not simple. The conditions for these phenomena
could not be recreated in a laboratory, and observations of them had
to be made over long periods of time at many stations in order to provide a complete record for study. The observers at the British colonial
stations in 1841 were in a unique position. They could measure the
intensity and direction of the terrestrial magnetic field at various
points across the world. Although each station realized that something was happening on September 25, it was not until months later
when the observations had been collected back in England, that scientists realized just how far-reaching the event was. Geophysical sciences, by their very nature, could not be studied in isolation or in
local areas alone. Only a systematic effort to collect data from a
worldwide series of points could provide the raw material for such
studies. The British colonial observatories, extending scientific observations into the British empire, were an important step in that
effort.

In recent decades, the historiography of imperial science has changed
considerably. Gone are the days when science was seen as merely "diffusing" from Europe into overseas possessions, as colonies struggled to
replicate the level of scientific institutions and techniques of their

---

"The Magnetic Crusade: Science and politics in early Victorian England."
Isis 70 (1979), 353. The earliest occurrences in the nineteenth century
appear to be from America in the 1840s. Farrar and Lovering refer to "the
present magnetic crusade" in their 1842 textbook on electromagnetism,
while Robert Patterson used the term in his 1843 address to the American
Philosophical Society. John Farrar and Joseph Lovering, *Electricity,
Magnetism and Electrodynamics* (Boston: Crocker & Ruggles, 1842), 243;
"Celebration of the Hundredth Anniversary, May 25, 1843," *Proceedings of
the American Philosophical Society* 3:27 (May 25–30, 1843), 34. I am
grateful to Marc Rothenberg for his assistance in finding these citations.

6. William Whewell, *History of the Inductive Sciences* (London: Parker & Son,
1857), 55.

home country.[7] Science has now become part of the discourse on imperial expansion, altering the traditional view of disinterested scientific activity. No longer "regarded as benevolent, apolitical, and value neutral," science has increasingly come to be viewed "as an agent of imperial ideology and enterprise."[8] Roy MacLeod integrates science into the culture of control and command, denying the objective position once ascribed to scientific research and replacing it with a political motivation toward the establishment of European imperial power.[9] Scientific discovery "produce[d] not only tools of conquest and occupation...but also instruments by which cultural and economic hegemony were secured."[10] Robert Stafford invokes Antonio Gransci's doctrine of hegemony to describe scientists "as agents and propagandists for the ruling class."[11] Lewis Pyenson concurs that French colonial scientists served as agents of French imperial interests.[12] Whether through collecting botanical and geological raw materials or physical and astronomical data, science served to further the "geopolitical interests of the European powers."[13]

Science extended European control in several ways. The first was the propagation of European language and culture. "Through science comes a language," notes MacLeod, and "through this language, neatly conveying the instrumental rationality of Western knowledge, comes control."[14] In the production of knowledge science also expanded imperial power. Suzanne Zeller argues that "as producers of useful knowledge," scientists were in a position to exercise "effective control over its explication and dissemination."[15] Control over such knowledge also enhanced the imperial ability to exploit and govern new territories. "Science has never been value-neutral or wholly objective; especially in the context of empire," argues Robert Stafford, "it implies control,

7. George Basalla, "The Spread of Western Science," *Science* 156 (May 1967): 611–622.
8. Roy MacLeod, "On Visiting the 'Moving Metropolis:' Reflections on the Architecture of Imperial Science,"*Scientific Colonialism* (Washington, DC: Smithsonian Institution Press, 1987), 218; MacLeod, "Passages in Imperial Science: From Empire to Commonwealth." *Journal of World History* 4:1 (1993), 120.
9. MacLeod, 1987, 243.
10. MacLeod, 1993, 118.
11. Robert Stafford, *Scientist of Empire* (Cambridge: Cambridge University Press, 1989), 189.
12. Lewis Pyenson, *Civilizing Mission* (Baltimore: Johns Hopkins University Press, 1993), 331. Hereafter Pyenson, 1993a.
13. MacLeod, 1993, 128; Pyenson, "Cultural Imperialism and Exact Sciences Revisited." *Isis* 84:1 (March, 1993), 104. Hereafter Pyenson, 1993b.
14. MacLeod, 1987, 218.
15. Suzanne Zeller, "The Colonial World as Geological Metaphor: Strata(gems) of Empire in Victorian Canada." *Osiris* 15 (2001), 105.

as an instrument both of administration and of knowledge."[16] Mapping and measurement were other methods of imperial expansion. Simon Schaffer sees the development of precision measurement by the nineteenth century as a sign of control over an organized society.[17] Deborah Warner expands the metaphor to empire: the "early modern period saw a proliferation of measurement corresponding with and undoubtedly attributable to the expansion of commerce and industry and increasingly, the imperial demands of the state."[18] Measuring the land and "mapping vast expanses of territory abroad" provided the opportunity "to annex new provinces to the domain of British science."[19] The expansion of science and empire went hand in hand. Observatories could play a role in this expansion of power, as physical points propagating both European cultural values and standards of measurement. Additionally, geophysical and astronomic observations facilitated commerce and the exploitation of natural resources in imperial possessions.

In this new view, however, science traditionally plays an acquiescent role; to use John Gascoigne's term, science in the service of empire. Here science is portrayed as an agent of imperial power, not a self-motivated force. Stafford describes this science as "sub-imperialism...meshing with the needs of the imperial government."[20] Science fits into an overall centralized view of empire, acting to expand power from the center and working primarily "for the greater glory of the imperialist power."[21] Scientific activity in the new territories is similarly relegated to a supporting role, serving the interests of the central science of the metropolis. MacLeod describes "a system of empirical science within a theoretical matrix determined from home."[22] In this program, centralized European science determined the topics of research and "the problems chosen as important were determined by the interests of the imperial power."[23] Pyenson views imperial scientists as mechanically filling in the gaps left by metropolitan science, fulfilling their role "as apostolic ambassadors instead of as independent, creative thinkers."[24] Since the scope of imperial science was determined at the center, there was no profit for "creating new research programs" at the

---

16. Stafford, 1989, 222–223.
17. Simon Schaffer, "Metrology, Metrication, and Victorian Values," *Victorian Science in Context* (Chicago: University of Chicago Press, 1997): 446.
18. Deborah Warner, "Terrestrial Magnetism: For the Glory of God and the Benefit of Mankind." *Osiris* 9 (1994), 69.
19. Stafford, 1989, 70.
20. Stafford, 1989, 223.
21. Pyenson, "Science and Imperialism," *Companion to the History of Modern Science* (London: Routledge, 1990): 927.
22. MacLeod, 1993, 129.
23. MacLeod, 1987, 244.
24. Pyenson, 1993a, 337.

periphery, thus colonial science consisted in "stewarding projects advanced by metropolitan mentors."[25]

However, such a centralized view of imperial science creates problems when applied to the specific situation of geophysical science in Britain during the nineteenth century. Unlike the later imperial structures of French, Dutch and German science, British science was remarkable decentralized through the empire.[26] The central controlling bodies were largely private institutions such as the Royal Society, not governmental or military ones. Additionally, there was little effort by the imperial state to establish a central authority on science. Only the admiralty served as an unofficial clearinghouse for scientific ventures such as geographic surveys. Many colonial scientists were left to their own devices when it came to research and founding observatories, museums and other scientific institutions.[27] The continued survival of colonial observatories often depended upon the willingness of colonial governments to continue funding them after British support had ended. Chambers and Gillespie question the applicability of the centralized metaphor for science, suggesting that "modern science is better understood both metaphorically and actually as a polycentric communications network."[28] In contrast to the centralized structures of continental science, Pyenson suggests that "there was little direct control of British colonial physics and astronomy from England. Physicists and astronomers in Canada and Australasia, while responding to commercial demands, followed their own karma in choosing research topics; their laboratories and observatories arose from local funding."[29] Stafford also admits the paucity of the "official [British] commitment to exploratory science...The continuing strength of the amateur natural history tradition...combined with the dominance of laissez-faire, dictated that nineteenth-century British science would be largely sponsored and managed on a voluntary basis."[30] Indeed, Stafford sees the activities of British geophysical scientists as an effort to "construct a global arena for self-fulfillment in the vacuum created by the imperial government's failure to establish a central department to manage official scientific enterprise."[31] These conditions provided an opportunity for various levels of scientific involvement, from observations made on board ships by naval officers to geomagnetic surveys conducted by amateur geologists.

---

25. Pyenson, 1993a, 331.
26. MacLeod, 1987, 237.
27. MacLeod, 1993, 129.
28. David Chambers and Richard Gillespie, "Locality in the History of Science: Colonial Science, Technoscience, and Indigenous Knowledge." *Osiris* 15 (2001): 223.
29. Pyenson, 1993b, 104.
30. Stafford, 1989, 205.
31. Stafford, 1989, 193.

Science on the periphery could have its own path, or influence work done at the center dramatically. Short of a centralized system of imperial aggrandizement, British science in the nineteenth century could be a mutually beneficial affair or even occasionally turned to the interests of science over those of empire. Stafford suggests that British science and empire could cooperate on equal terms. "While the scientists won access to widening career options and new data, the imperial government gained accurate information for administering and developing its sprawling possessions."[32] Pyenson cites examples when imperial science pursued pure research rather than applied, seeking answers to "general questions in astronomy and physics that would return little immediate profit to metropolitan centers."[33] John MacKenzie concludes "the relationship of western science with the rest of the world is not simply a matter of 'relocation' or 'diffusion,' but also a learning process in which peripheral discoveries created revolutions."[34] MacLeod leaves the question open. "It remains to us to discover more instances in which the institutions and leadership of Britain were dependent upon colonial discovery and enterprise."[35]

The purpose of this book will be to examine the interactions of science and empire, not from the standpoint of centralized science in the service of empire, but through the mutual influence that center and periphery could have on one another. The specific context of geophysics in the nineteenth century allows us to explore cases when the central science of the metropolis was indeed dependent upon the work done at the periphery, an example especially heightened by the particular relationship between geophysics and the British empire. At the same time, the politics of an imperial setting could influence both the content and style of science; the type of observations that could be made and the way those observations were shaped into theories. In this reciprocal relationship, empire and science worked in mutual assistance.[36]

Three separate trends came together in the early nineteenth century, allowing for the first worldwide research into scientific phenomena. The first prerequisite was European imperial expansion, which had brought Europe into contact with new lands and territories stretching around the globe. The second trend, built upon the first, was a renewed interest in the study of geophysics. Topics such as terrestrial magnetism and meteorology could now be studied on an appropriate scale

---

32. Stafford, 1989, 189.
33. Pyenson, 1993a, 332.
34. John MacKenzie, "Introduction," in *Imperialism and the Natural World* (Manchester: Manchester University Press, 1990): 3–4.
35. MacLeod, 1987, 245.
36. Michael Worboys, "The British Association and Empire: Science and Social Imperialism, 1880–1940," *The Parliament of Science* (Northwood: Science Reviews Ltd., 1981): 171.

and there was the possibility of reducing these erratic phenomena to simple scientific laws. The global setting was necessary for geophysical science because these fields could only be studied by observation, not *a priori*. Inherently empirical, they were not experimental; geomagnetic conditions could only be experienced in the field, not recreated in the laboratory.[37] Any unifying force behind them could only be found through induction.

The final trend, growing cut of the first two, was an interest in universal science, a search for a way to ground inductive research on a foundation as firm as mathematics. Geophysics and induction shared similar limitations when it came to expanding particular observations to universal laws. Just as geophysical observations were valid only for the time and place they were taken, so inductive data only applied to the specific cases involved. Neither could be extended beyond those local conditions to produce a general conclusion. (Rain one day did not guarantee rain the next, nor would the compass always point due north.) Geophysics could also only be studied through an inductive method that was decentralized. Regular observations had to be made at points around the world, requiring an extensive number of observers and observing points. No one scientist or even small group of intellects could hope to do all of the necessary work. Thus the empire made an ideal staging ground for geophysical research, if science could employ the resources of empire to its ends.

The conditions that created the opportunity for global geophysical study came together in Britain during the early nineteenth century. A revival of geophysical interest and Baconian methods suggested the possibility of a universal approach to inductive science, one that could eliminate the particular nature of the observations and give a general necessity to the results. The three trends reinforced one another. Geophysics could only be studied through the scope provided by empire, and only understood through the universalization of induction. One could not study geophysical events in isolation or individually; limits of space and time had to be overcome. British or European surveys alone could never provide the necessary observations to construct a worldwide theory. Thus imperial observing points could expand the scale of observations, allowing the center to act at a distance, as Bruno Latour has put it. Expeditions and observatories "mobilized anything that can be made to move," be it botanical specimens or observations of geomagnetic variations, "and shipped [them] back home for this universal census."[38] Yet this massive collection of information still left the problem of linking the unconnected observations into a cohesive theory. What was to be done with this data overload when it reached

37. Warner, 68.
38. Bruno Latour, *Science in Action* (Cambridge: Cambridge University Press, 1987): 225–228.

the "center of calculation?"[39] Here universal science, influenced by its imperial setting, could finally develop.

John Herschel was a well-known astronomer who participated in all three of these nineteenth-century trends. His trip to the Cape Colony in the 1830s for the purpose of observing the skies of the southern hemisphere introduced him to the importance of observations at the periphery and the question of validating universal inductive conclusions. Although a colonial observer, Herschel still enjoyed a somewhat privileged position in astronomy, as the heavens obligingly moved, allowing him to sweep the entire hemisphere from one position. Herschel began to realize how much more time and effort would have to be put into geophysical observations that could only be made at multiple points around the globe. Yet the possibilities inherent in such observations were apparent, and the imperial setting of the Cape, thousands of miles from the scientific metropolis of London, greatly influenced Herschel's thought on the matter.[40]

Once he returned to England, Herschel's interest turned to geophysics, where he discovered the same problems existed when it came to the issues of observation and induction. No matter how many observations he made, he could never get a complete picture. There were always gaps left to be filled in. Thomas Hankins describes how the problem was especially acute with the orbits of double stars, which Herschel studied closely as it was difficult to determine if binary stars were real or apparent. In many instances, there were only scattered observations that could be used to construct the orbit of a binary system. With centuries of positions left unobserved, how could he determine the true path of the stars? Herschel solved the problem through graphical analysis, using a curve to approximate the orbit of the star by using the best fit to its known positions and extrapolating the missing points. He later realized that such a graphical method would also work for geophysical observations.[41] Peter Barlow's magnetic research had already shown that "geometrical techniques could be applied to the problem of constructing representations of magnetic phenomena," thus making empirically derived curves "respectable" to the mathematically inclined.[42]

---

39. Latour, 232–233.
40. Similarly, Elizabeth Green Musselman argues that the "unfamiliar social and political" setting of the Cape inspired Herschel's belief in using science and empire for rational improvement in civilization. Elizabeth Green Musselman, "Swords into ploughshares: John Herschel's progressive view of astronomical and imperial governance," *British Journal for the History of Science* 31:4 (December 1998): 420.
41. Thomas Hankins, "A 'Large and Graceful Sinuosity': John Herschel's Graphical Method." *Isis* 97:4 (December 2006), 606
42. David Gooding, "'Magnetic Curves' and the Magnetic Field: Experiment and Representation in the History of a Theory," *The Uses of Experiment* (Cambridge: Cambridge University Press, 1989), 207.

Herschel eventually incorporated his graphical treatment of induction into his overall philosophy of science, which provided a basis for geophysical research in the British empire. Whether it was finding the magnitude of geomagnetic activity or elements of double star orbits, he needed observations.[43] Thus Herschel joined an effort in the late 1830s to assemble a network of imperial observatories that could provide the observational data needed to allow universal inductive conclusions to be drawn in geophysics. These observatories produced the necessary raw material for the development of a theory at the center, but without these peripheral listening posts, the center had nothing on which to build. Even dozens of stations, however, were limited in the amount of the globe they could cover or the number of years they could operate. Here inductive universalism inspired by graphical analysis provided an answer. "Herschel recognized that graphs provided a practical method for dealing with complex data. They allowed the natural philosopher to smooth out random errors of observations and fluctuations in the phenomena being observed. They also exposed order that would be concealed in a table of numbers."[44] By filling in the gaps left by observation, Herschel could create a pattern in geophysical phenomena just as in stellar obits. By extending this method, universal results, valid for places and times there had not been (or had not yet been) observations could be obtained. Thus the joint potential of colonial observations and the functionality of universal induction through graphical representation, both suggested by the new imperial setting, allowed geophysical research on a scale that would have been impossible before.

When I began my research, I was interested in Herschel and the role that he played in the study of geomagnetism and particularly the Magnetic Crusade in the nineteenth century. From the existing scholarship, I was led to believe that Herschel's role had been largely secondary to the main scientific work done by others.[45] However, as I studied Herschel's correspondence and his activities in the years leading up to the crusade, I discovered that he had been working along similar lines for some time. Indeed, the crusade appeared to be the culmination of a line of observational research that led back to his own colonial adventures in southern Africa. Herschel believed in a universal induction that required multiple observations of a phenomenon across time and space to achieve certainty. He thought that a worldwide system of observatories could provide the basis for scientific theories of terrestrial magnetism, meteorology and the tides.

---

43. Hankins, 624.
44. Hankins, 606.
45. Jack Morrell & Arnold Thackray give an excellent account of the history of the Magnetic Crusade in *Gentlemen of Science* (Oxford: Clarendon Press, 1981): 353–370.

Using Herschel as a starting point, I could trace the contributions that he had made, not just to the lobby for the crusade but also to the central ideas behind the venture. Privately, or through personal friends, Herschel had been trying to accomplish the same goals as the crusade for years. The factors that came together in the late 1830s finally gave him his opportunity.

Herschel thus became a critical part of the lobby for the crusade. In concert with other scientists who had an interest in geomagnetism, and with the nascent British Association for the Advancement of Science, this lobby pressed the British government to support their scientific project. During the course of the lobby, the crusade itself underwent changes in planning. I realized that the form of the project was being shaped not just by the scientific ideas of its founders, but also to appeal to different factions within government to avoid particular departments where a difficult reception could be expected. In short, the very political process through which the scientific project gained approval helped to shape the final form of that project. The expedition which launched in 1839 reflected not only the original ideas of John Herschel, Edward Sabine, Humphrey Lloyd and other scientists who had backed it, but also the contributions from members of government, admiralty and Royal Navy.

I was interested to see if a comparable process of political influence would occur in a similar scientific project in this period. Fortunately, the United States also launched an Antarctic expedition (commonly known as the Wilkes expedition) in the 1830s. By comparing the process the American project had to go through to gain governmental approval, I was able to see the same political influences at work, although in a different setting. Just as the British proposal had to adjust to accommodate the requirements of its political system, so the American proposal had to be altered to ensure its very survival in the competitive political climate of the antebellum age. In some ways, the American political environment made it even more difficult to succeed and hence had a greater impact on the final form of the project. Indeed, many elements of the scientific proposal that appealed to the aristocratic British government were damaging, if not fatal, in the American system. In the end, both governments approved and sent out an expedition to the southern hemisphere. The differences in those expeditions and the steps that were taken to follow up on their discoveries reproduce many of the differences in the sociopolitical systems that created them.

The theme that tied all of my research together was imperial science. Both Britain and American were using imperial exploration and expansion as the staging ground for science in the nineteenth century. Whether overseas or through western domestic expansion, each country was studying the world on a larger scale than could have been accomplished without such an imperial structure. Venturing out into the world, science found itself enmeshed in newly forming empires. Both science and the state had much to gain in cooperating in this new imperial field. Science provided the technical knowledge for expanding the empire and making it profitable. The empire provided the grounds on which science could set up its observatories and the stations from which they could observe. Behind both, the state supported scientific ventures financially and established a growing empire. This cooperation between state and science, between imperialism and empiricism, is at the heart of this book.

My argument is that the British empire provided the necessary resources for the creation of a universal inductive geoscience that was shaped by the political and social realities of the state apparatus that sponsored it. Universal science would have been impossible without the imperial setting and the resources of empire that influenced not only the collection of data, but also the theory behind it. The two sides, state and science, played mutually reinforcing roles. Without imperial resources, science could never have gathered the information it needed to construct a cohesive theory. Without universal science, the data would be no more than a static collection of curiosities, frozen in time and useless for imperial exploitation. Each side needed the abilities and assets of the other in order to produce useful results. Accordingly, the final result reflected the mixture of imperial and scientific capital that went into the project.

The establishment of geophysical observatories marked a key intersection in this process. Britain had sponsored expeditions for years that made limited observations of natural phenomena such as geomagnetism and meteorology. However, these temporary snapshots of the effects of the earth's magnetic field or an approaching storm front could not reveal the regular patterns needed for these fields to develop. A series of consistent observations over a longer span of time was needed. The physical observatories required for such observations necessitated a larger investment of resources, both of time and money, on the part of the British state. Thus the scientific venture could not help but become enmeshed in the imperial arena. Rather than imperial interests seizing upon such stations to further the scope of British power abroad, the advocates of science had to convince the state of the benefits and advantages of such outposts. Herschel, representing the scientific side of the crusade, often found himself at odds with his primary ally, Edward Sabine, an officer of the Royal Artillery who was more accustomed to the temporary scientific surveys carried out as a secondary goal by military expeditions. The two initially placed a different emphasis on the purpose of the crusade, Herschel seeing it as a chance to establish a worldwide system of observatories and Sabine viewing it primarily as a military/scientific research voyage. Over the course of the lobby, Herschel was able to win over Sabine as well as the state to his belief in the importance of fixed observatories. This support was not an open-ended commitment, however. As the years passed, supporters of the crusade fought to legitimize its goals and maintain its funding in the face of public doubts and tightening budgets. In the long run, the same combination of forces that had brought the crusade into existence could hamper its continued success.

The Magnetic Crusade was the creation of a number of factors, all of which came together in 1839: A revival of empiricism, influenced by the work of Alexander von Humboldt, the extension of British influence in the world and the participation of several key scientific figures all played their part. At the center were John Frederick William Herschel, England's premier astronomer and Edward Sabine, an artillery officer with an interest in all things scientific. Both were interested in the issues raised by geophysical science and realized that research into such fields could only be accomplished with the aid of imperial resources. Yet the process worked both ways. While Herschel and Sabine were agents of empire, they also employed the empire for their own ends. Their respective social positions allowed them to frame their scientific interests in terms appealing to the state in order to achieve their goals. Repeatedly, both would have to rely upon their own assets to accomplish their task. In the end, empire both provided a setting where universal science could be practiced and influenced the shape of scientific enterprises, helping to overcome the inherited problems of inductive science and to set up a method by which scientists could begin to study interconnected phenomena on a global scale.

Chapter One

A Fitting Enterprise
of a Maritime People

Before science and empire could work together on a worldwide
scale, there had to be the opportunity for interaction between science and state. Until the nineteenth century, such prospects were limited. With the exception of occasional voyages of discovery, most
states were unwilling to invest in scientific ventures unless there was
some immediate profit or strategic goal. The expanding imperial world
opened up new possibilities for contact between science and state. The
geophysical sciences especially were of interest, as they were based in
and tied directly to the growth of empire. One of the first major projects embarked upon by imperial science involved geomagnetism, a
field of prime interest to both empiricism and maritime expansion.

SCIENCE AND STATE

Traditionally, the British government had been reluctant to
involve itself with the expense of sponsoring scientific ventures.
International investment was more likely to go into trading companies
that yielded profits and acquired new territories. In the years before the
Magnetic Crusade, cooperation between science and the British state
was sporadic at best. Nevertheless, several such enterprises had been
financed by Great Britain. Not surprisingly, most of them had to do

with the sea. In the 1690s, the British government had sent explorers (including Edmund Halley) to seek Terrae Incognitae Australis, the unknown southern continent.[1] While Halley's voyages failed to locate the continent, they did provide the opportunity to collect measurements on the variation of the compass from geographic north in the Atlantic.[2] These data allowed him to print the first detailed map of magnetic declination (or isogonic chart) in 1701.[3] His voyage also convinced him that geomagnetism, winds and ocean currents had to be studied on a global scale.[4] In the 1770s Captain James Cook circumnavigated Antarctica without sighting it. Not until the early nineteenth century did British and American sailors, seeking whales and seals, make their way through the ice-covered seas to the outer edge of the Antarctic continent. Although the official discovery came in the 1830s, attempts to locate this extreme southern land continued to fuel new scientific and navigational endeavors.

Both the voyages of Halley and Cook contributed to discoveries of a different kind. Ever since William Gilbert had described the Earth as a magnet in 1600, the question of the location of the Earth's magnetic poles had remained. For maritime European societies that still used magnetic compasses as their primary navigational instrument, the operation of the Earth's magnetic field was of prime concern. By the early nineteenth century, the location of an Arctic magnetic pole had been established. In subsequent years, the search for the Antarctic pole drew ships further and further south. No private individual or enterprise was capable of equipping and dispatching such ventures. Explorers and scientific enthusiasts had to turn to state institutions for funding and support. However, not all governments were willing to employ public money on such projects. The history of state-sponsored science before the nineteenth century was sporadic at best. Even nations with long-standing maritime traditions were not easily convinced. British involvement in state-sanctioned science provides an example of the limited results that could be expected.

The travels of Captain Cook in the eighteenth century set the pattern for other voyages of discovery which involved scientific observers, including Darwin's trip on the *Beagle* in the 1830s. Cook's first voyage had been to set up an observing station at Tahiti to observe the transit

---

1. DNB.
2. Halley also set up an astronomic observatory on St. Helena. He was already interested in geomagnetism and had published his own theory of terrestrial magnetism fifteen years earlier. Edmund Halley, "A Theory of the Variation of the Magnetical Compass." *Philosophical Transactions*, 13 (1683), 208–221.
3. Edmund Halley, *The Description and Uses of a New and Correct Sea Chart of the Whole World* (London: Hartigan, 1701).
4. G. E. Fogg, "The Royal Society and the Antarctic." *Notes and Records of the Royal Society of London*, 54:1 (January 2000), 85.

of Venus in 1769 as part of a scientific venture which involved British and continental scientists in a cooperative attempt to set up a series of observations around the world in time for this rare event. Occurring in pairs that came only every 150 years, the transits provided an excellent method of measuring solar parallax, from which the distance from the Earth to the sun could be calculated.[5] The Royal Society took the lead in planning the observations and in soliciting funding from the government. Lord Macclesfield, president of the Royal Society, employed the rhetoric of nationalism by declaring that foreign governments "expected" Britain to participate in the worldwide effort.[6] In its appeal to the treasury for financial support, the Royal Society warned that the French had already committed to the project and appealed to British pride, claiming that the transits had been first predicted by an Englishman (Halley) and had only been observed "but once before since the world began" by another Englishman (Horrox).[7] Eventually, the Royal Society was successful in gaining government aid for this scientific venture. In the eighteenth and nineteenth centuries, the Society became a key bridge between science and state in Britain.[8]

On later voyages, Cook mapped large portions of the Pacific and collected thousands of plant and animal specimens. Cook's efforts served both science and the state. In addition to charting (and claiming) new lands for Britain, he brought new flora and fauna to the attention of science and participated in astronomical and geophysical observations. William Goetzmann calls these voyages "the epitome of Baconian or Lockean empirical observation conducted over a huge portion of the globe."[9] Cook's voyages can be seen as the beginning of the British connection of private empirical and public imperial interests which expanded in the nineteenth century. Lucile Brockway has claimed that "private and government enterprise cannot be rigidly separated in colonial expansion...Cook and Sir Joseph Banks, using a

5. See Harry Woolf, *The Transits of Venus* (Princeton: Princeton University Press, 1959).
6. Macclesfield to Duke of Newcastle, July 5, 1760. (APS Film #808).
7. Royal Society to Commissioners of the Treasury, July 9, 1760. (APS Film #808).
8. John Gascoigne argues that the Royal Society employed itself in the interest of the British state in part due to the common gentry composition of the society and the British ruling class. He suggests that the gentry values of the Royal Society encouraged state service and promoted mercantilist and defensive policies. Thus the Royal Society and its leadership become deeply involved in imperial affairs, offering scientific advice to foster the growth of the empire. The expanding empire provided commercial and military security, as well as opportunities for new science, thus satisfying both the gentry's obligations to the state and its scientific curiosity. See John Gascoigne, *Science in the Service of Empire* (Cambridge, Cambridge University Press, 1998), 111–146.
9. William Goetzmann, *New Lands, New Men* (New York: Viking, 1986), 52.

combination of Royal Society money and private money (Banks'), explored, botanized (bringing back 17,000 plant specimens), observed the transit of Venus, visited the Society Islands which Cook named after the Royal Society, and acquired Australia for the Crown, all in one voyage."[10]

By far most expansive British investment in science concerned attempts to correctly determine longitude. As late as the mid-eighteenth century, there was no reliable way of determining longitude onboard ship.[11] The discovery of such a method would have been invaluable to seafaring nations dependent on navigation for commerce and defense. In 1714, Parliament set up an institution known as the Board of Longitude to encourage research and to consider applications for government aid. Parliament offered £20,000 for anyone who could come up with a successful method of determining the longitude of a place to within thirty nautical miles. Parliament also authorized the board to give out smaller grants for promising inventions and applications that might lead to determining longitude or to other improvements in navigation.[12] A successful method would not only provide a way to measure longitude on land, but also on a ship at sea.[13] Additionally, in the process of finding a solution, the search had the potential to stimulate other fields of science such as terrestrial magnetism, cartography and geodesy.

Applicants tried numerous approaches to the problem of longitude. Astronomical attempts focused on finding a way to predict the movement of the moon against the stars, which could be used as a timekeeper. Earth-bound efforts concentrated on everything from detecting variations in the Earth's magnetic field to developing more accurate methods of dead reckoning (estimating the mileage traveled by ships). While John Harrison's invention of the naval chronometer eventually proved a reliable solution to the problem, the board continued to exist into the early nineteenth century, and provided one of the few official sources of government patronage for science in that period.[14]

In the aftermath of the Napoleonic Wars, there was a new push for government retrenchment or belt-tightening. In 1816 the income tax

---

10. Lucille Brockway, *Science and Colonial Expansion* (New York: Academic Press, 1979), 188.
11. Galileo's discovery of the Jovian moons in the seventeenth century provided a method for finding longitude provided that careful telescopic observations could be made.
12. A. R. T. Jonkers, *Earth's Magnetism in the Age of Sail* (Baltimore: Johns Hopkins Press, 2003), 30.
13. For the search for longitude see Dava Sobel, *Longitude* (New York: Walker, 1995).
14. It was not until 1765 that the board approved Harrison's innovation, and several decades more before the navy finally adopted it. Jonkers, 144.

had been abolished, depriving the state of much-needed revenue. In the years after Napoleon, the British debt reached £900,000,000. To try to compensate for this shortfall, government expenditures were reduced significantly during the 1820s.[15] Consequently, less financial support was generally made available for science.[16] One field that still found some hope of support was that of geophysics which, due to its connection with the question of longitude, could still appeal to the board for aid. As early as the seventeenth century, Henry Bond had claimed that longitude could be found by comparing the difference between true north and magnetic north at a particular location on the globe.[17] Up until the nineteenth century, many attempts were made to refine his method, using the variation of the compass needle away from geographic north to find the longitude.

However, the board was frugal with its gifts. As one of the few places scientists or inventors could turn for government assistance, it received many applications every year. The board rejected most of them, such as one J. Nicholl's method of ascertaining the variations of the magnetic needle in 1804 or a proposal for the construction of a chart of the variation of the magnetic needle by a Mr. Yeates in 1816.[18] Few of the proposed innovations offered a better method of determining the longitude or even much accuracy in doing so. A "magnetic wheel" offered by one Mr. Wood in 1818 was not considered to be useful; it made the longitude of Portsmouth to be five degrees east of London![19] When it did consider geomagnetic proposals, the board was interested in successful applications, not theories. In 1815 Lieutenant Edward Naylor of the Royal Marines submitted a paper on the theory of the variation of the compass, which the board ruled to be out of their field of interest.[20] Later attempts to provide a successful application for this theory were also rejected. Naylor's computations of the declination of the compass based on his system were off by two or three degrees and it was finally judged that his system was not useful.[21]

There were however, some applicants who met with success. In late 1794 Ralph Walker had invented a "meridinal" compass that he

---

15. Robert K. Webb, *Modern England* (New York: Harper & Row, 1980), 157–172.
16. Polar exploration, however, expanded due to the availability of navy ships after the end of hostilities. Fogg, 86.
17. Henry Bond, *The Longitude Found* (London: Godbid, 1676).
18. Board of Longitude Minutes, March 1, 1804; March 7, 1816. (PRO RGO 14/7–8). Hereafter cited as Minutes. One has to wonder, however, if the board missed a golden opportunity in rejecting a "Mr. Stowe's discovery of Perpetual Motion, [for] having nothing to do with the finding of the Longitude"! Minutes December 4, 1817.
19. Minutes, February 4, 1818. (PRO RGO 14/3–4).
20. Minutes, December 7, 1815. (PRO RGO 14/7–8).
21. Minutes, February 3, 1820. (PRO RGO 14/3–4).

claimed could compensate for magnetic variation and thus find the longitude at sea.[22] He wrote to claim the full reward of £20,000. The board, after consulting with the astronomer royal, questioned whether any method of detecting magnetic variation could actually provide a useful means of determining the longitude. Nevertheless, there was sufficient interest generated by Walker's invention for the board to order further tests of his compass.[23] The report on the compass came back, finding that it did indeed have some use in determining magnetic variation. The board then decided to subsidize Walker to develop his idea further, purchasing one of his compasses and ordering that he be paid the sum of £200 as a reward for the improvement.[24] Eventually the British navy adopted Walker's compass design.[25]

Another man who received support from the Board of Longitude was Captain Edward Sabine. Sabine (1788–1873) was an artillery officer educated at Woolwich. He fought in the War of 1812 at Fort Eire and eventually reached the rank of major-general in 1859. Although the military was his "proper profession," science was Sabine's lifelong pursuit.[26] He was involved in investigating several branches of geophysics by conducting research on the shape of the Earth and terrestrial magnetism. He also participated in a number of Arctic voyages. He became a fellow of the Royal Society in 1818 and although he did not join the British Association when it was first founded, he later became a very active member and served as its general secretary from 1839 until 1859. He also served as president of the Royal Society from 1861 to 1871 and was knighted in 1869.[27]

In 1821 the Royal Society sent Sabine to the vicinity of the equator to make observations that could help in determining the shape of the Earth.[28] The technique involved using a pendulum to determine the varying gravitational pull at different places on the Earth's surface, which indicated the length of the radius of the Earth at various points and thus its overall shape.[29] As had been realized in the eighteenth century, the Earth is not a perfect sphere, but rather bulges at the equator and flattens at the poles. This was an effect predicted by Newton and confirmed by various expeditions that measured the Earth's radius at polar and equatorial points. Because of the bulge, the

---

22. Walker was from Jamaica and had written *A Treatise on Magnetism*, published in London in 1794. Jonkers, 120.
23. Minutes, December 6, 1794. (PRO RGO 14/6).
24. Minutes, June 6, 1795. (PRO RGO 14/6).
25. Jonkers, 149.
26. Sabine to Hansteen, October 10, 1826. (ITA).
27. DNB.
28. DSB.
29. Edward Sabine, "An Account of Experiments to Determine the Acceleration of the Pendulum in Different Latitudes." *Philosophical Transactions* 111 (1821): 163–190.

Earth's radius is slightly longer at the equator and the acceleration due to gravity is consequently slightly smaller. This bulge also affects the length of a degree of longitude.[30] The board considered this expedition in geodesy worthwhile, and resolved to furnish Sabine with any instruments that he might require.[31] Later, Sir Humphrey Davy forwarded a letter from Sabine to the board, proposing to continue the experiments. Again, the board agreed to support Sabine.[32] For his pendulum research, Sabine later won the Copley medal, the highest award given by the Royal Society.

Sabine, however, did not always enjoy the favor of the board. In 1825, following his return from the tropics, he made a new proposal for a series of meteorological observations on the peak of Tenerife in the Canary Islands. Sabine appealed to John Herschel, then a member of the board who had earlier complimented Sabine on his pendulum work, predicting that his geodetic results would cause an "extraordinary sensation."[33] Herschel and Sabine had already worked together on a joint Anglo-French project to determine the exact difference in longitude between Greenwich and Paris. Herschel felt that Sabine's new project was a desirable one, and he hoped that the board would approve it, as he believed that Sabine was only requesting funding to convey himself and his equipment to the Canaries.[34] However, he did not envision the board taking on quite the obligation that Sabine wished.

Sabine wanted the observations to receive full and official government support. "The experiments in question either are or are not a public object; either are or are not worthy of being executed at the public expense" he declared. Sabine realized that a extensive series of observations required the support of more than just one committed individual. He saw the backing of the Board of Longitude as a way of assuring that such a project would be carried out and be taken seriously. He felt that the execution of his experiments was certain if authorized and supported officially by the board, but only contingent if left to his individual resources.[35] Herschel, however, had doubts that the board would be willing to undertake the full expense of the project.[36]

Sabine, learning that the board would not back his mission to Tenerife, was disappointed. He was prepared to undertake the full expense himself, but lacked the means to do so. While Sabine was

---

30. See Goetzmann, 19–25.
31. Minutes, June 7, 1821. (PRO RGO 14/7–8).
32. Minutes, November 7, 1822. (PRO RGO 14/7–8).
33. Herschel to Sabine, March 20, 1825. (RS HS 15.1). An invaluable reference for Herschel's correspondence is *A Calendar of the Correspondence of Sir John Herschel*, Michael Crowe, ed. (Cambridge: Cambridge University Press, 1998).
34. Herschel to Sabine, April 3, 1825. (RS HS 15.2).
35. Sabine to Herschel, April 4, 1825. (RS HS 15.3).
36. Herschel to Sabine, April 4, 1825. (RS HS 15.4).

grateful for Herschel's personal support, he required official sanction, not just private goodwill.[37] Herschel could only suggest that Sabine look instead for aid from the Royal Society or the admiralty.[38] At present, the means did not exist to arrange state support of a scientific venture on such a scale as Sabine proposed. In the future, both Herschel and Sabine put themselves in a position to extend the range of state aid for science. Indeed, the two men later combined their efforts on a far more extensive project in the field of geomagnetism than that which Sabine had just proposed.

The precedent of Cook had established the state's willingness to become involved in scientific activity and indicated the possible link between science and empire which later developed. However, in the early nineteenth century the British government had yet to commit itself fully to scientific goals except those that lay within its interest as a maritime and commercial power. Government funding was hard to find for science and few official agencies existed to provide an avenue for scientific proposals to reach the state. By 1830 the Board of Longitude itself had been dissolved, eliminating one of the few sources of government patronage.[39] Following the demise of the board, the Royal Navy became the chief source of scientific patronage, but it was not always reliable.[40] Scientific ventures could be postponed or even canceled if relations between science and admiralty became strained.[41] Only scientists with political or social connections within the government were consistently able to extract pecuniary support.

GEOMAGNETISM

Geomagnetic interest in the nineteenth century was inspired primarily by the work of Alexander von Humboldt. In 1799, von Humboldt had set out on an expedition to study the natural

---

37. Sabine to Herschel, April 6, 1825. (RS HS 15.5).
38. Herschel to Sabine, April 8, 1825. (RS HS 15.6).
39. Herschel was so disconcerted by the dissolution of the board that he refused to collect the £75 still owed to him as a commissioner of longitude, "having ceased to consider myself a public functionary from the moment I became acquainted with the intentions of government respecting the dissolution of the Board of Longitude." Herschel to Morton, January 12, 1830. (RS HS 21.54). Fortunately for Herschel's finances, he received the commission from Lardner to write the *Preliminary Discourse* (along with a promise of £250) just one week later. Lardner to Herschel, January 19, 1830. (RS HS 11.115).
40. Alfred Friendly, *Beaufort of the Admiralty* (New York: Random House, 1977), 289.
41. Sabine to Hansteen, July 3, 1826. (ITA)

world in the Spanish territories in Latin America. Here he took measurements of the geomagnetic elements (among other things) and developed his idea that magnetic force decreased from pole to equator. He also believed that geomagnetism (along with all other terrestrial phenomena) was influenced by "cosmic" forces coming from outside of the Earth. Humboldt's magnum opus, *Kosmos*, was an attempt to demonstrate that a unified field connected the entire terrestrial system.[42] After his return to Europe, Humboldt campaigned for the establishment of geomagnetic observatories for further study. Numerous continental observatories were set up for this purpose, especially in Russia and in Germany where a union (Verein) of observatories was established by 1834 by the physicists Carl Gauss (recruited by Humboldt in 1818) and Wilhelm Weber.[43] Weber's work at the observatory at Goettingen contributed to Gauss's new theory of geomagnetism, which attempted to measure magnetism in Newtonian terms.[44]

Britain was also interested in geomagnetism. It had been realized for centuries that understanding the variations in the Earth's magnetic field was crucial for ships reliant upon magnetic compasses. Most scientific investigation of geomagnetism originated in states trading and expanding overseas.[45] Thus it was not uncommon to see military voyages and explorations making magnetic observations. The origins of the Magnetic Crusade may be traced to one such journey, an expedition into Arctic waters in 1818 whose crew included James Ross and Edward Sabine.[46] Exploring the Northwest Passage, they made a series of magnetic observations on the ice as well as onboard ship. Although the magnetic poles were known to lie in the Arctic and Antarctic regions, it was difficult to reach them even if they could be precisely located. Sabine regretted that they had been unable to make more

---

42. Goetzmann, 59.
43. S. R. C. Malin and D. R. Barraclough, "Humboldt and the Earth's Magnetic Field." *Quarterly Journal of the Royal Astronomical Society* 32 (1991), 279–293. By 1835, geomagnetic stations were operating at Altona, Augsburg, Berlin, Breda, Breslau, Copenhagen, Dublin, Freiburg, Goettingen, Greenwich, Hanover, Leipzig, Marburg, Milan, Munich, St. Petersburg, Stockholm and Upsala. William Whewell, *History of the Inductive Sciences* (London: John Parker, 1857), III:50.
44. Robert Multhauf and Gregory Good, *A Brief History of Geomagnetism* (Washington DC: Smithsonian Institution Press, 1987), 16.
45. Warner, Deborah, "Terrestrial Magnetism: For the Glory of God and the Benefit of Mankind." *Osiris* 9 (1994), 83.
46. Ross (1800–1862) had entered the navy at the age of twelve and was actively engaged in polar exploration for much of his life. Ross became a fellow of the Royal Society in 1828 and discovered the north magnetic pole in 1831. He received post rank in 1834 and was later involved in the Magnetic Crusade. DNB.

observations, being closer to the predicted location of the north magnetic pole than ever before.[47]

Sabine later claimed that it was during the time of this voyage that
he and Ross made plans to eventually carry out magnetic observations
on a massive scale, an intention only fulfilled years later. Writing to
Herschel in 1840, Sabine recalled what he termed the "private history"
of the Magnetic Crusade:

> Scene, Baffins Bay. Dram[atis] Pers[onae], Sabine & Ross, observing the
> Dip, 88 ½, at which time & place it was agreed that when the N. W. pas-
> sage should be made, & Ross should be a port Captain, Sabine should
> recommend & Ross execute a magnetic survey of the globe, commencing
> with the southern hemisphere, with 2 ships, etc. etc. just as we now
> have it. The subject has never dropped with me, tho' my earlier efforts,
> some of which you know, fell very dead, and such I believe could have
> been the fate to this day, but for the Brit[ish]. Ass[ociatio]n.[48]

There was no lack of activity on geomagnetism during the 1820s and
1830s. In 1822, Sabine commented on the increased attention that had
been given to the subject of magnetism in recent years and expressed
the desire that scientists should strive for a greater degree of accuracy
in observing its various terrestrial phenomena.[49] Sabine himself went
on two voyages in 1822–1823 where he made magnetic observations in
the north Atlantic.[50] But the Norwegian astronomer Christopher
Hansteen carried out the major geomagnetic work done in the 1820s.
Hansteen was influenced by the German universal idea of
Naturphilosophie, which led him to see a unified view of nature.[51] For
him, geomagnetism was only part of a cosmic magnetic force that did
not arise entirely from within the Earth—a view that he shared with
Alexander von Humboldt. Thus the sun, moon, or other extraterrestrial forces could influence geomagnetism.

In geomagnetism, Hansteen followed the earlier work of Edmund
Halley, who had allowed for four or more magnetic poles (at least two
north and two south), representing the axes of two concentric spheres
within the Earth to explain terrestrial magnetism. Hansteen had won
the 1811 prize offered by the Royal Society of Science at Copenhagen
for his thesis that it was necessary to employ more than a single

47. Sabine, "On Irregularities Observed in the Direction of the Compass
    Needle." *Philosophical Transactions* 109 (1819), 121–122.
48. Sabine to Herschel, July 6, 1840. (PRO BJ3/26/135).
49. Edward Sabine, "An Account of Experiments to Determine the Amount of
    the Dip." *Philosophical Transactions* 112 (1822), 1.
50. Edward Sabine, "Report on the Variation of the Magnetic Intensity."
    BAAS Report 7 (1837), 11.
51. John Cawood, "Terrestrial Magnetism and the Development of
    International Collaboration in the Early Nineteenth Century." *Annals of
    Science* 34, (1977), 568.

magnetic axis in order to explain the magnetic phenomena on Earth.[52] Hansteen's belief, like Halley, was that the Earth in fact had two magnetic axes, "one stronger, the other weaker."[53] He denied the possibility of a single axis because observational evidence showed that the magnetic dip was not directly proportional to the intensity at corresponding latitudes.[54] The main problem in determining a geomagnetic theory was that the Earth's magnetic field itself seemed to vary, causing the magnetic elements to change over time. Halley had explained this by allowing his two magnetic axes to slowly rotate, thus changing the positions of the magnetic poles at either end of the axes. As a result, Hansteen's theory required four magnetic poles or "points of convergence," two in each hemisphere.[55] He believed this conclusion was "as evidently demonstrated as any of the best founded propositions in our actual physical features."[56] After working out his own theory on the nature of the Earth's magnetic field, Hansteen set out to test it by crossing Russia making magnetic observations. In 1828 he left St. Petersburg and eventually reached Irkutsk, finding evidence for a magnetic pole in Siberia.[57] Sabine and Hansteen had been in correspondence since 1825, and Sabine looked forward to the results of Hansteen's Russian expedition, which Sabine referred to as "your most meritorious undertaking."[58] The two shared a belief in Halley's theory that there were four magnetic poles, two in each hemisphere, caused by two magnetic axes of unequal strength.[59]

In later years, Sabine became the chief proponent of Hansteen's theory in Britain, even bringing out an English abstract of Hansteen's "Magnetismus der Erde" in 1835.[60] Hansteen's aim was to verify his theory by locating the four poles and the magnetic equator (or balance point) between them. To this end, he desired additional observations of dip and intensity in Liberia, Terra del Feugo, and near the equator in the Indian and Pacific Oceans.[61] Given the limited resources of his home country, Hansteen hoped Sabine might be able to obtain British assistance, believing that "if the English Government will not give a generous and powerful hand to researchers...there is little hope that the theory shall advance."[62] In 1828 Hansteen was glad to hear of a

52. Edward Sabine, "On the Phenomena of Terrestrial Magnetism." *BAAS Report* 5 (1835), 62.
53. Ibid., 61.
54. Ibid., 66.
55. Ibid., 69.
56. Hansteen to Sabine, May 13, 1826. (PRO BJ3/3/13).
57. Sabine, 1837, 23.
58. Sabine to Hansteen, March 29, 1828. (ITA).
59. Hansteen to Sabine, February 16, 1826. (PRO BJ3/3/11).
60. Sabine, 1835, 61–90.
61. Sabine to Hansteen, July 4, 1827. (ITA).
62. Hansteen to Sabine, May 13, 1826. (PRO BJ3/3/13).

new expedition being sent to the equatorial regions under Captain Henry Foster on behalf of the Royal Society, believing that it would provide the "materials necessary to the foundation of a perfect establishment of the elements of magnetical theory of the Earth."[63] Foster's voyages took him throughout South America making observations, but these were apparently not enough to finish Hansteen's work. The next year Hansteen wrote back to Sabine that he still required observations around Australia and along the western coast of South America, as well as observations onboard ships.[64]

With the work of Ross in the Canadian Arctic and Hansteen in Siberia during the 1820s, knowledge of the magnetic field in the northern hemisphere seemed fairly complete. In 1830 Sabine was comfortable enough with the success in the north to turn his attention to the south. "If we now direct our attention to the southern hemisphere, we find nearly the whole field of enquiry untrodden." He suggested that a single scientific voyage could cover much ground and be fairly inexpensive.[65] This was the beginning of British plans for an Antarctic geomagnetic voyage that led to several abortive attempts before the success of the Magnetic Crusade.

In 1833, under the auspices of the annual report of the British Association, S. Hunter Christie published the "State of Our Knowledge Respecting Geomagnetism" as part of the British Association's attempt to present the most up-to-date theories for every field of science. Christie (1784–1865) had taken his degree at Cambridge in 1805. He served for almost fifty years as a professor at the Royal Military Academy at Woolwich, where Sabine also served. He became a fellow of the Royal Society in 1826 and was known for his interest in the magnetic sciences.[66] In his report, Christie outlined the basics of magnetic theory. There were three measurable elements in geomagnetism: the magnetic variation, dip and intensity.[67] Variation represented the

63. Hansteen to Sabine, April 20, 1828. (PRO BJ3/3/19). Foster, like Sabine, was a military man with an interest in science. He had been elected a fellow of the Royal Society in 1824 after he had traveled with Sabine on an Arctic expedition. In 1826 his magnetic and astronomic observations were printed in the *Philosophical Transactions* at the expense of the Board of Longitude. He won the Copley medal in 1827. Foster spent three years on his equatorial mission but did not survive to return to Britain, unfortunately drowning in Panama. Henry Foster, "Corrections to the Reductions of Lieutenant Foster's Observations on Atmospherical Refractions at Port Bowen; with Addenda to the Tables of Magnetic Intensities at the Same Place." *Philosophical Transactions* 117 (1827): 122–125. DNB.
64. Hansteen to Sabine, April 18, 1829. (PRO BJ3/3/28).
65. Edward Sabine, "Observations on the Magnetism of the Earth." *American Journal of Science and Arts* 17:1 (1830), 152–153.
66. DNB.
67. The terms for these elements vary, sometimes confusingly so. Variation is often referred to as declination, while dip is called inclination. I will use

amount by which a compass deviated from due north as measured by the pole star. The variation was of great concern to sailors, as it represented the amount that the compass needle varied from true geographic north, a difference that could mean significant course corrections. Christie maintained that this phenomenon had been known since 1269, and many nautical charts had attempted to build in a compensation factor.[68] Such adjustments had to be constantly updated, based on new observations. Any factor included on a chart was, at best, valid only for the period in which the chart was printed. Within a few years, a new factor had to be added to compensate for the additional variation. Dip was the vertical "dip" of the compass needle below the horizontal plane, which some believed to be related to latitude. Finally, intensity represented the strength of the magnetic field at a given place, measured by how long a compass needle set in motion by a magnet required to stabilize. As these elements themselves varied over time, they had defied all attempts to reduce them to a single regular theory. Such a theory was desirable, as it would not only give a clue as to the nature of the Earth's magnetic field, but also because the ability to predict the future secular changes of the magnetic elements would greatly aid navigation.[69]

According to Christie, there was general agreement that geomagnetic theory could not properly be determined without magnetic observations from various positions around the world.[70] The required observations could either be conducted on a naval expedition or through established stations. The time seemed right for a venture of this nature in the 1830s.[71] Europe was at peace and international conflicts did not appear to be a threat.[72] In 1838 Herschel wrote to

---

the terms employed by Christie. Christopher Carter, "Magnetic Compass," *Oxford Companion to World Exploration* (New York: Oxford University Press, 2007), 204–205.

68. Humboldt asserted that the Chinese had recognized the effect of magnetic variation in the Song dynasty, "nearly four hundred years before Christopher Columbus and the natives of Europe had the least idea of magnetic declination." Alexander von Humboldt, "On the Advancement of the Knowledge of Terrestrial Magnetism, by the Establishment of Magnetic Stations and Corresponding Observations." *London and Edinburgh Philosophical Magazine* 9 (1836), 46.

69. Sabine, 1835, 105–130.

70. Samuel Hunter Christie, "Report on the State of our Knowledge Respecting the Magnetism of the Earth." *BAAS Report* 3 (1833), 130.

71. Giuliano Pancaldi, "Scientific Internationalism and the British Association," *The Parliament of Science* (Northwood: Science Reviews Ltd., 1981), 156.

72. Peace and war could have significant effects on science. In the eighteenth century, the Seven Years War had threatened the success of the joint efforts to observe the transits of Venus by fueling Anglo-French animosity. In the nineteenth century, the Crimean War interrupted the period of

Francis Baily "if peace continues and the world goes on for another quarter century as it is doing now, we shall know something!"[73] In 1834, Sabine heard about an effort being made by the French physicist Francois Arago to bring to the British government's attention the importance of founding observatories in its colonies.[74] Arago's proposal was the first suggestion of using British colonies as observing posts for fixed stations. Sabine, however, still favored an expedition.

By the 1830s, recent discoveries in the field of geomagnetism had brought it to a point where an experimentum crucis was needed to settle the central split between the two-pole and four-pole theories. The geomagnetic discoveries of Hansteen and Ross suggested that the Earth's magnetic field had two poles in the northern hemisphere, backing the argument that the Earth possessed more than one magnetic axis. However, Carl Friedrich Gauss, in his *Allgemeine Theorie Des Erdmagnetismus* (1841) argued for one axis and two poles. Gauss had solved the geomagnetic theory as if it were a function $V(r, q, j)$ that represented the irregular distribution of magnetism throughout the Earth. The partial differentials of V described the magnetic field at any particular point. Gauss's accomplishment was made possible by earlier observations of the three geomagnetic elements that were necessary to complete the theory. However, the constant coefficients of the equation V could only be solved through direct observation of the magnetic field at a particular point.[75] Gauss's theory required only two magnetic poles. In addition, Gauss believed that geomagnetism was an internal (not cosmic) force, causing him to look for a solely terrestrial source for magnetism.[76]

The British scientific community itself took sides in this debate. Herschel adopted Gauss's two-pole view of geomagnetism as well as

scientific cooperation that started after the Congress of Vienna in the 1850s. Only after the Congress of Berlin in 1878 did the international situation again become conducive to scientific expeditions.

73. Herschel to Baily, April 7, 1838. (RS HS 25.8.10).
74. Lloyd to Sabine, September 23, 1834. (PRO BJ3/7/37).
75. G. D. Garland, "The Contribution of Carl Friedrich Gauss to Geomagnetism." *Historia Mathematica* 6 (1979), 14; Whewell, *History Inductive Sciences*, III:53. Gauss in particular cited the work done by Sabine during the period 1835-1838 for contributing to his theory. John Herschel, "Terrestrial Magnetism." *Quarterly Review* 66 (1840), 285.
76. Garland, 18. Hansteen and Gauss were actually working on different magnetic phenomena. Hansteen's poles were based on readings of magnetic intensity, while Gauss's poles were those based on inclination (the primary magnetic poles). Gauss's theory did actually predict the existence of four regions of high intensity in approximately the regions cited by Hansteen. Whewell, 1857, III:52.

its terrestrial origin.[77] Sabine was in agreement with Hansteen.[78] The effort to solve the problem of geomagnetism strongly influenced the course of events. For the Halley/Hansteen side the discovery of the two southern poles was imperative in order to determine the positions and strengths of the magnetic axes. For the Gaussian side, the location of the pole was less important than a series of observation that could provide the needed data to calculate the unknown coefficients of Gauss's formula. In either case, British science alone could not accomplish the necessary goals. Only by turning to the state could they find a source of support and influence that allowed them to conduct science on a global scale.

By 1835, British plans for an Antarctic voyage were underway. Enthusiasm for a geomagnetic project produced what Sabine called a "magnetic fever" in the British Association.[79] This activity can be seen as the first magnetic lobby. Within the British Association a group including Sabine, Ross and the Reverend Humphrey Lloyd began to press the British government to support such an expedition. Sabine's 1835 publication "On the Phenomena of Terrestrial Magnetism" put forth Hansteen's view of the magnetic field complete with its two southern poles (the two northern poles having been located in Canada and Siberia by Ross and Hansteen respectively). Convinced that four poles explained the Earth's magnetic phenomena, Sabine called for an expedition to the south to find the remaining two poles.[80] Sabine also asked for Hansteen's advice on the proposed route of the expedition.[81] This effort seems to have had some support. Writing in 1840, Sabine recalled that his 1835 report to the British Association was meant to be a proposal to the Government for a global magnetic survey. The effect of the report:

> was to produce much discussion of the subject at that meeting; & of the *impression* produced...that an Antarctic expedition, such as we have now in all respects except as regards the Magnometers, was *all* but recommended at that meeting; & four port Captains amongst them Sir J. Franklin...volunteered their services for it. Several of the principle

77. Herschel also resisted cosmic explanations in meteorology. He once commented to Schumacher that "it is singular how generally prevalent the opinion is that both my father and myself have advocated the idea of the Moon's influence on the weather and published predictions and weather tables founded on such presumed influence. But there is not the shadow of foundation for such idea—on the contrary, all possible pain has been taken on the part of both him and myself to disavow and disclaim all such pretended tables and predictions." Herschel to Schumacher, March 1841. (RS HS 22.101).
78. Sabine to Lloyd, June 19, 1836. (RS Te #16). Ross was another supporter of Hansteen's theory. Ross to Sabine, October 10, 1838. (PRO BJ3/16/16).
79. Sabine to Herschel, July 6, 1840. (PRO BJ3/26/135).
80. Sabine, 1835, 90.
81. Sabine to Hansteen, August 19, 1835. (PRO BJ3/82/11).

members of the Association pledged themselves to the support of the subject.[82]

However, this first effort met with little success. Although Hansteen enthusiastically supported this new venture, it did not generate enough support to have a chance of successfully lobbying the Government.[83] Having already approved a coastal survey of North America, the British government was unwilling to take on another major scientific project the same year.[84] Additionally, news that the American government was planning a similar voyage (the Wilkes expedition) dulled British interest.[85] William Whewell wrote in 1836 that nothing had been done that year respecting the Antarctic magnetic expedition. He even reported that support for the expedition within the scientific community was not as strong as it could have been. Whewell hoped that next year, with Sabine's report on the matter in circulation, there would be a better chance for action.[86] But by the time of the Bristol meeting of the British Association in 1836, Lloyd lamented that nothing had been done about the Antarctic expedition, and feared that it would be neglected for some years.[87] Sabine, though, had hopes that the project was "only deferred, not defunct."[88]

Despite this disappointment, geomagnetism received a boost in 1836 with an appeal from Alexander von Humboldt to the Duke of Sussex, then president of the Royal Society. Humboldt, already involved in European magnetic observations, believed that British overseas possessions could be used as observation points for geomagnetism. He pressed the Royal Society to help establish observing stations in the colonies to promote the advancement of the mathematical and physical sciences.[89] Humboldt's letter was referred to committee but not acted upon.[90] Humboldt also appealed to Sabine for assistance in

82. Sabine to Herschel, July 7, 1840. (PRO BJ3/26/136).
83. *"Le prepos honorable de marquer la route d'une expedition magnetique autour du pole antaritique m'a 'ete' extremement agreable."* Hansteen to Sabine, January 26, 1836. (PRO BJ3/3/31).
84. Sabine to Hansteen, August 1, 1836. (ITA).
85. Jack Morrell and Arnold Thackray, *Gentlemen of Science.* (Oxford: Clarendon Press, 1981), 355–356. Indeed, the approval of an American Antarctic expedition that same year caused a delay in similar British efforts, although later news of the success of the Wilkes expedition also had a salutary effect on the British scientific lobby. See Chapter IV.
86. Whewell to Harcourt, May 27, 1836. (WC).
87. Lloyd to Sabine, September 1, 1836. (PRO BJ3/7/84).
88. Sabine to Hansteen, November 14, 1836. (ITA)
89. Humboldt, 42–53. Specifically, Humboldt suggested that stations be established in New Holland (Australia), Ceylon, Mauritius, the Cape ("rendered illustrious by the labours of Sir John Herschel"), St. Helena, and North America.
90. Morrell and Thackray, 357.

this proposal, writing to him about the instruments and stations that he wished the Royal Society to adopt.[91] However, Sabine failed to become involved, perhaps because the plan did not focus on an expedition.[92] Still, Lloyd hoped to occasionally take part in the observations now going on throughout Europe.[93]

In spite of this failure, domestic research continued. In 1833 Lloyd had begun a magnetic survey of Ireland, about which he sought the advice of James Forbes in Scotland.[94] This project continued with the help of Sabine, whose artillery division had been relocated to Dublin in 1830 as a result of the disturbances in Ireland that accompanied Parliament's debate of Catholic Emancipation. Humphrey Lloyd (1800–1881) became involved in a number of geomagnetic projects; he was once called "the British Oracle on this subject."[95] Educated at Trinity College (Dublin) he became a professor there in 1831, succeeding his father, Bartholomew Lloyd. He was the technical advisor for geomagnetic observations, and trained many of the observers who worked in the colonial stations. Lloyd served as president of the Royal Irish Academy from 1846 to 1851 and was provost at Trinity College from 1867 to 1881.[96] During the 1830s Sabine and Lloyd conducted a full magnetic survey of the British Isles. While Sabine covered Scotland, Lloyd traveled through Liverpool, Shrewsbury, Hertford, Chiseton, Bristol, Salisbury, Ryde (on the Isle of Wight), Brighton, London, Cambridge, Lynn, Matlock, and Manchester.[97] In 1837, Lloyd and Sabine formed "a quadruple alliance" with Ross and John Phillips (a Welsh geologist) to cover the remainder of Britain.[98] Sabine was occupied for much of 1837 working on a report of magnetic force lines that he had hoped to publish in the fall of 1836. The delay had been caused by the return of Captain Robert Fitzroy from the South Seas with six years' worth of observations that Sabine felt he should include.[99] In 1838, Lloyd took over the magnetic observatory at Dublin

---

91. Sabine commented to Lloyd "of course he [Humboldt] supposes, that both as an Ordnance officer, & as an old secretary of the R S, who has written several papers on magnetic subjects in the Transactions, that I have something to say in the affair." Sabine to Lloyd, February 9, 1837. (RS Te #26).
92. Some historians have suggested that Sabine prompted Humboldt to write in 1836, although Sabine's failure to follow through casts doubt on this possibility. John Cawood, "Comments," *Human Implications of Scientific Advance.* (Edinburgh: Edinburgh University Press, 1978), 145.
93. Lloyd to Sabine, February 9, 1836. (PRO BJ3/7/78).
94. Sabine to Lloyd, August 21, 1833. (RS Te #1); Lloyd to Forbes, August 20, 1833. (SAUL msdep7, Incoming Letters 1833 #39).
95. Forbes to Lloyd, March 19, 1840. (SAUL msdep7, Letterbook III pp.84–86).
96. Morrell and Thackray, 536.
97. Lloyd to Sabine, October 31, 1836. (PRO BJ3/7/86).
98. Lloyd to Sabine, August 3, 1837. (PRO BJ3/7/104).
99. Sabine to Phillips. August 1, 1837. (APS 509 En3).

founded by his father, proposing to devote his time to studying magnetic variations and theorizing about them.[100] International geomagnetic research continued with more expeditions in the north equipped to make magnetic observations.

As interest in geomagnetism spread, science began to engage the imperial structure. While local parties could handle domestic surveys, expanding these observations abroad required the support of British imperial institutions, such as the East India Company. In early 1838, Captain Thomas Jervis, an officer with the East India Company, suggested a magnetic survey in India and central Asia. Lloyd wrote to Sabine that he had received a letter from Jervis requesting suggestions as to what should be done in India in the field of geomagnetism. Lloyd believed (inaccurately) "that his inquiries relate chiefly to the course to be pursued at fixed magnetic stations."[101] The limited resources available to British science could not carry out such an extensive project, forcing Jervis to turn to his employers. He made the proposal through the Royal Geographical Society to the court of directors of the East India Company. Acknowledging their role in geographic exploration in the past, Jervis proposed that the company outfit an Asian expedition with scientific instruments. He challenged the court to do so lest Britain be overtaken by the continental powers (and especially Russia) in this field.[102] Jervis received support from the Magnetic Committee of the Royal Society (formed to consider Humboldt's 1836 request) that recommended at least two observatories in India.[103] The East India Company was less enthusiastic though, and rejected Jervis's appeal a few months later.[104] However, Jervis renewed his efforts later that year.

Sabine meanwhile kept up the pressure and continued to push Hansteen's view of geomagnetism. His 1837 "Report on the Variation of the Magnetic Intensity Observed at Different Points on the Earth's Surface" reiterated his support for Hansteen's four-pole hypothesis and the need for an expedition to locate the two southern poles. Sabine held that a geomagnetic theory could not be completed until "observations in the vicinity of the magnetic poles themselves" were obtained and that knowing the location of the poles was necessary to show lines of equal magnetic intensity, declination and variation on maps. Such

---

100. Lloyd to Herschel, June 9, 1838. (RS HS 11.263).
101. Lloyd to Sabine, February 18, 1838. (PRO BJ3/8/19).
102. Jervis to E.I.C., February 21, 1838. (BL Add 34649 fol. 83).
103. Magnetic Committee to Lords Commissioners of the Admiralty, April 20, 1838. (RS Te #56).
104. E.I.C. to Jervis, July 1838. (BL Add 34649 fol. 87). One possible reason for the Company's lack of interest was that the early stages of the first Anglo-Afghan War had just begun in the spring of 1838, no doubt diverting Company resources away from any scientific pursuits in the area. Brian Gardner, *The East India Company.* (New York: Dorset Press, 1971), 209.

graphical analysis was important to the study of geomagnetism because it was impossible to observe every location on the surface of the Earth. Knowledge of the poles and general observations at key points allowed information for the rest of the globe to be extrapolated. Sabine reminded his readers that England had been "the first country which sent an expedition expressly for magnetic observation, namely, that of Halley...it is right that we should feel a particular interest in that in which we have ourselves led the way." He declared that the most important service that Britain could render to this branch of science was in "filling the void which still existed in the southern hemisphere." His rhetoric indicated that Sabine had not given up on the idea of a southern expedition and that he would welcome another magnetic lobby, "a fitting enterprise of a maritime people."[105]

Thus the years leading up to 1838 had seen several unsuccessful attempts at a British geomagnetic project on a global scale: Arago's suggestion of using colonial observatories, Sabine's Antarctic expeditions, Humboldt's colonial stations and Jervis's Indian survey. When the second lobby came about, it was more ambitious; elements from all of these ideas were combined. The new crusade involved both a southern expedition and colonial stations. The new effort used the British Association rather than the Royal Society as its primary vehicle to lobby the government and drew upon many of the same people who had been involved in the first lobby. A significant change was the involvement of John Herschel.

Edward Sabine and Humphrey Lloyd, two prominent members of the first unsuccessful lobby, decided to try again in the fall of 1838. Most recently, Sabine and Lloyd had worked together on a magnetic survey of the British Isles. In order to increase the chances of success of a new lobby, they decided to appeal to John Herschel for his support. Recently returned from Africa, where he had been observing the southern hemisphere skies for the last five years, Herschel seemed to be the ideal choice to join the lobby. Herschel was an aristocratic scientist with social connections as well as an interest in the field of geomagnetism. Earlier in the year, Herschel had feared that the government had given up on a magnetic voyage or any other magnetic undertaking.[106] Now Sabine and Lloyd were delighted to find Herschel had indicated interest in the subject of a new lobby.[107] Lloyd also knew that Herschel wished to set up British magnetic observatories in correspondence with the German Verein of Gauss and Weber.[108] Support for a new geo-

---

105. Sabine, 1837, 82–85.
106. Herschel to Sabine, January 30, 1838. (PRO BJ3/26/1–2).
107. Sabine to Phillips, July 4, 1838. Reprinted in *Gentlemen of Science: Early Correspondence of the BAAS*, Jack Morrell and Arnold Thackray, editors (London: Royal Historical Society, 1984), 263.
108. Lloyd to Sabine, June 26, 1838. (PRO BJ3/8/40).

magnetic undertaking appeared to be building. "There is something quite inspiring in the sounds that issue from the cave of the cyclops," wrote Lloyd to Sabine, anticipating Herschel's cooperation.

> The interest is not lessened by the knowledge that the arms are forging there for those we know & value & for a campaign which is likely to be successful & glorious. Herschel's advocacy of the South Polar expedition will, I think, almost insure its progress.[109]

Sabine and Lloyd planned to bring up their venture at the next meeting of the British Association, in Newcastle. Before that time, Herschel's support had to be secured. In order to appeal to his known sympathy for fixed observatories, stations were included in the new plan in addition to the revised expedition. Later, Sabine claimed that his plan for the crusade had always encompassed both elements of an expedition and fixed observatories:

> The magnetic members of the Association always contemplated that both branches should be cared for...My visit to Ireland, previous to the Newcastle meeting, was expressly to represent to Lloyd & Robinson that it was plain the machinery in that quarter could not work; & to propose that our plan should comprehend both branches, employing Artillery officers & mil[itary] officers for the observatories.[110]

Lloyd had agreed to this approach, and together Sabine and Lloyd decided to approach Herschel for his support for a new magnetic lobby. Thus on August 4, 1838, the push for the Magnetic Crusade began. The first goal of the lobby was not to be securing government support, but the support and cooperation of John Herschel.

---

109. Lloyd to Sabine, July 2, 1838. (PRO BJ3/8/42).
110. Sabine to Herschel, July 6, 1840. (PRO BJ3/26/135).

Chapter Two

The Knowledge of Many...
Attainable by One

Philosophical beliefs have often had a great impact on scientific style. The thought of Bacon, Locke, Hume and Kant had all contributed to the methods of doing empirical science in the seventeenth and eighteenth centuries. Although influenced and modified by social conditions and realities, at its heart the science of Sir John Herschel was also philosophical. However, he certainly was not acting in an idealistic vacuum. Herschel's inductive approach to science was designed to meet philosophical problems raised during the eighteenth century, and his method was tailored to the expanding worlds of British Empire and science in the nineteenth century. That his philosophy was so functional in these changing times is a sign of its effectiveness. Herschel always held to his belief in a universal approach to science, both due to his Baconian heritage and the contemporary needs of the field. His inductive solutions provided new ways of dealing with the problems faced by scientists in the imperial period. Herschel's science found its place in a growing world of new measurements, specimens and phenomena that had not been imagined a century before. With Herschel's aid, universal science and British colonialism, empiricism and imperialism, meshed in a cooperative venture.

⬛ JOHN HERSCHEL

John Frederick William Herschel was born March 7, 1792, the only son of famed astronomer Sir William Herschel (1738–1822). Much of Herschel's social position and influence he owed to his father. William Herschel had immigrated to England from the family home in Hanover in 1757 and initially made his living as a musician. By 1773 however, he had found his passion: telescopes. His new interest led him to build larger and more accurate scopes and his observations using these instruments led to the discovery of the planet Uranus, the first new planet known since antiquity, in 1781. Subsequently, William Herschel became court astronomer to George III. He was knighted in 1816 and had a comfortable income of £200 a year from his royal post.[1] John Herschel grew up in a family of astronomers; in addition to his father, his aunt Caroline Herschel was a noted observer. The Herschel family was also a prominent gentry one, but one that had gained prestige through scientific, not military, service. This background put Herschel in an almost unique position in the early nineteenth century. His family wealth and accolades connected him to Britain's ruling class, while his father's reputation gave him a natural position in the field of science.[2] As a kind of "aristocratic" scientist able to fund his own interests, Herschel was in a position to contribute to science in ways that others could not.

Herschel's education started at home. After a brief time at Eton in 1800, he was tutored privately until his departure for Cambridge in 1809, where he studied mathematics and physics.[3] After submitting a mathematical paper to the Royal Society in 1812, he was elected Fellow of the Royal Society in 1813, the same year he completed his B.A. in mathematics.[4] Despite an early inclination toward law, he returned to Cambridge to take his M.A. in 1816 and became involved in astronomy.[5] In 1821 and 1824 he toured the continent meeting various scientists including Arago, Laplace, Biot, Humboldt, Poisson, Fourier

---

1. DNB.
2. At his death, William Herschel left £25,000 to his son, along with property in Buckinghamshire and scientific equipment. However, it is not clear where all of William's money came from, as his post as astronomer to the court of George III paid only £200 a year. *London Times*, October 9, 1822 2:d; DSB.
3. Günther Buttman, *The Shadow of the Telescope* (New York: Charles Scribner's Sons, 1970), 9–10, 13.
4. John Herschel "On a Remarkable Application of Cotes's Theorem." *Philosophical Transactions* 103 (1813): 8–26. Herschel won the Royal Society's Copley medal in 1821 and 1847 and the Royal Medal in 1833, 1836 and 1840. DSB.
5. Buttman, 15, 19.

and Encke.[6] He was knighted in 1831 (into the same order as his father)—the year after he composed his philosophical treatise on science, the *Preliminary Discourse on the Study of Natural Philosophy*.[7] His career in astronomical observing was marked by the five years he spent at the Cape of Good Hope (1833–1838) surveying the southern hemisphere. Here Herschel began to appreciate the potential of colonial observatories.[8] After his return from Africa in 1838 he received a baronetcy. It was at this time that he became involved in the project later known as the Magnetic Crusade.

Herschel's influence in science during the second quarter of the nineteenth century is difficult to overestimate. Susan Faye Cannon declared that to be a scientist in this period was "to be like John Herschel."[9] Her definition was apt. Herschel was often seen as the model of a scientist in the nineteenth century. In its 1871 obituary, the *London Times* remarked that he had secured "the widest reputation among men of science, both at home and abroad; while his numerous popular writings have contributed largely to the diffusion of a taste for science."[10] Charles Darwin listed Herschel's *Preliminary Discourse* as one of the works that most influenced him.[11] John Stuart Mill and Michael Faraday were others who learned from (and admired) Herschel. Herschel had the advantage of living in a formative age for science.[12] Empiricism, idealism, religion and politics all contributed to the formation of modern science in the early nineteenth century.

Herschel became involved in many fields of science, perhaps too many. His attention might be frantically focused upon one field for a time, as long as a couple of years. But then he would be off on a new

---

6. DSB.
7. Joseph Agassi is correct in stating that the Preliminary Discourse was published in 1831 (despite the date on its title page), but the book was clearly written in 1830. Lardner sent proof copies to Herschel in October 1830 and had the corrections back by December. Joseph Agassi, "Sir John Herschel's Philosophy of Success." *Historical Studies in the Physical Sciences*, 1 (1969), 1; Lardner to Herschel, October 2, 1830. (RS HS 11.119); Lardner to Herschel, December 6, 1830. (RS HS 11.120).
8. Steven Ruskin argues that the British government used Herschel's trip to the Cape in the 1830s to "legitimize its control and possession of the Cape Colony in southern Africa," although the Cape had been in British hands since 1806 and had legally become a British possession in 1814. Ruskin also neglects the presence of Thomas Maclear, who was astronomer at the Cape observatory even before Herschel arrived. Steven Ruskin, *John Herschel's Cape Voyage* (Burlington: Ashgate, 2004), 38, 69.
9. Susan Faye Cannon, *Science in Culture* (New York: Dawson, 1978), 36. Cannon's treatment of Herschel provides the basis of this thesis and many other later studies.
10. *London Times*, May 13, 1871, 5:5.
11. Cannon, 1978, 55.
12. Ibid., 225.

track, abandoning his previous research or leaving it to others to develop. Herschel himself once confessed to having "too many hankerings."[13] His interests can be seen across the fields covered during his long career: astronomy, optics, meteorology, geomagnetism and photography. Perhaps because of this tendency to become involved in so many areas, his contributions have often been overlooked, even in fields where they were crucial.[14]

NINETEENTH CENTURY SCIENCE

By the early nineteenth century in Britain, science was expanding so much that it was difficult for old assumptions to keep up. In 1831, for example, William Whewell noted that magnetism had advanced so quickly in the early nineteenth century, that it had accomplished as much as astronomy had "from the first contemplation of the moving skies by Chaldean shepherds, to the demonstration of universal gravitation by Newton."[15] Throughout the eighteenth century, science had been "Newtonianized," brought into the Newtonian synthesis to be explained by force laws or treated experimentally. By the nineteenth century, fields of science, which Newton had barely commented upon, now had to be brought under the heading of universal Newtonian science; however, in fields such as optics, Newtonian conclusions were challenged. As science became more specialized, it fragmented. The notion that there was some universal science or some general way to treat the sciences was challenged.[16] The rapid increase in scientific facts and knowledge had, by the early nineteenth century, threatened to break apart the discipline and to eliminate the unity of science which held that a chemist and an astronomer (or a geologist and a physician) were somehow pursuing the same end. It became increasingly difficult to define exactly what this field of "science" involved. Also, increasing specialization made it difficult if not impossible for a single individual to keep up with, much less make significant contributions in, multiple fields. The days of polymaths seemed to be rapidly ending.

Adding to these disturbing trends was the belief by many British scientists that science in their country was in decline by the early nineteenth century.[17] "The state of science in England! Lord help us,"

---

13. Herschel to Whewell, July 22/23, 1837. (RS HS 21.224).
14. Cannon, 1978, 245.
15. [Whewell], "Herschel's Preliminary Discourse." *Quarterly Review* (1831,)395. This was an anonymous review but outside evidence indicates Whewell as the author. Hereafter cited as [Whewell].
16. Agassi, 2.
17. See Morrell and Thackray, *Gentlemen of Science* (Oxford: Clarendon Press, 1981), 47–58.

Herschel exclaimed to Charles Babbage in 1829. "We are a conceited nation and have our ignorance to learn."[18] Babbage's *Reflections on the Decline of Science in England* (1830) represented the views of many scientists in the country.[19] Herschel held that in many great branches of science Britain was no longer the preeminent nation.[20] "In mathematics we have long since drawn the rein, and given over a hopeless race. In chemistry the case is not better...There are, indeed, few sciences which would not furnish matter for similar remark."[21] Herschel thought that in all fields Britain now lagged behind the other European countries. Part of the impetus for geomagnetic research in Britain was the perception of "British backwardness compared with Continental achievements."[22] Herschel even felt that there was not a good English scientific journal being published.[23] He felt that the French *Annales de Chemie* far surpassed that "crude and undigested Scientific matter which suffices (we are ashamed to say) for the monthly and quarterly amusement of our own countrymen."[24] Whewell later declared British science in this period to be a "great empire falling to pieces."[25]

At the same time, some scientists attacked the failings of the Royal Society, which by the nineteenth century had become more of a social club than a scientific institute, its membership composed of noble amateurs and opportunists.[26] Francis Beaufort complained of the "venal motives" of many members who "endeavour to increase their professional practice, or give *éclat* to some silly book by imposing the appendage to their names of FRS."[27] Herschel and Babbage also shared a negative view of the proceedings of the society.[28] Some called for a

---

18. Herschel to Babbage, December 15, 1829. (RS HS 2.242).
19. Although Herschel feared that Babbage's blunt approach "would have done much more had it been less bitterly sarcastic." Herschel to Babbage, May 22, 1830. (RS HS 2.252).
20. Herschel to Whewell, February 15, 1831. (RS HS 21.79).
21. Quoted in L. Pearce Williams, *Michael Faraday* (New York: Basic Books, 1965), 350.
22. David Miller, "The Revival of the Physical Sciences in Britain: 1815–1840." *Osiris*, 2 (1986), 128.
23. "John Herschel Letter." *South African Libraries*, 7 (1940): 138-154. Hereafter cited as Herschel, 1834.
24. Quoted in Williams, 349.
25. William Whewell, *Philosophy of the Inductive Sciences* (London: John Parker, 1847) II:4.
26. See Morrell and Thackray, 52–58.
27. Quoted in Friendly, 285–286. Beaufort (1774–1857) was appointed royal hydrographer in 1829 and served in that position for twenty-six years. DNB.
28. Herschel once wrote to Francis Baily, apologizing for his absence at an astronomy committee meeting because he had not been able to "overcome that repugnance as to venture my nose into what Babbage calls pandemonium." Herschel to Baily, December 19, 1830. (RS HS 21.76).

complete reform of science and its various institutions. "This is an age of reforms," wrote Herschel (appropriately) in 1830. "I hope and trust that no false step on the part of the Royal Society will prevent its benefiting as much as possible from the present conjunction."[29]

In an effort to push reform on the Royal Society, some members nominated Herschel for the presidency in 1830.[30] Herschel, however, had qualms about accepting such a post. "I wish the R[oyal] S[ociety] a better President than I should make it," he wrote to Babbage. Herschel claimed that he "love[d] science too well to be very easily induced to throw away the small part of the lifetime I have to bestow on it on the affairs of a public body which has proved to me ever since I became connected with it a continued source of disappointment and annoyance."[31] In the face of Herschel's noncandidacy, the society elected the Duke of Sussex as president, angering some reformers who sought to restructure the scientific establishment, even if it meant breaking away from the Royal Society to do it.

Partially as a result of the reform movement in 1830, a number of scientists formed the British Association for the Advancement of Science (BAAS) as a professional alternative to the Royal Society. Many of the talented young scientists involved in the redefinition of science were members of this new association, although there were also many crossovers who belonged to both the Royal Society and the British Association. Surprisingly Herschel, the former candidate of the reformers, was less than enthusiastic about the new venture. A firm believer in the individual nature of scientific discovery, Herschel had little use for large societies.[32] He saw perfect spontaneous freedom of thought as the essence of scientific progress, and believed that the only use of people combining into societies was to do what individuals could not do for themselves alone. Herschel feared that a large professional society could stifle the pursuit of individual scientific initiative by involving too many competing (and misinformed) ideas. "A thousand bad opinions do not make one good one—a thousand mediocrities don't make one excellence, nor can a thousand eyes looking through as many spy glasses see as well or as far as one with a first rate telescope."[33] Yet despite his misgivings, Herschel became an important member of the new association, and also became involved in one of its great projects in the 1830s.[34]

29. Herschel to Babbage, May 22, 1830. (RS HS 2.252).
30. Cannon, 1978, 170–173.
31. Herschel to Babbage, October 15, 1830. (RS HS 2.255).
32. In contrast to Herschel's misgivings about the British Association, Forbes believed that "the [British] Association is itself the best Meteorological Society that ever was formed & capable of doing almost anything." Forbes to Herschel, October 25, 1838. (SAUL msdep7 Letterbook II pp. 233–236).
33. Herschel to Whewell, September 20, 1831. (RS HS 25.2.21).
34. David Miller connects the reform of science and the revival of the physical

Much has been written about the conflict between the "aristocratic" Royal Society and the new "bourgeois" British Association in this period.[35] By the early nineteenth century the traditional gentry lifestyle had been challenged by bourgeois values with the growth of industry and the rise of the middle class. Some historians have seen this gentry/ bourgeois divide in science as well, both in the scientific method and in the competition between the aristocratic Royal Society and the new middle class societies, such as the Astronomic and Geographic Societies. However, in spite of this logical divide the split between the Royal Society and British Association was often more illusion than real during this period. Indeed, men such as Herschel became a key link between the two bodies, turning to the Royal Society for assistance when he realized that the lobby was in danger of failure. Responsibility for the lobby became so mixed that at times it becomes difficult to tell which body was doing what. Their joint aim was clear: to utilize the resources of the British state and its expanding imperial possessions to further the needs of universal science. While the Royal Society provided a more direct connection with the aristocratic (and imperial) interests of the ruling class, the British Association was also interested in imperial expansion insofar as it provided locales for observations. During the lobby for the crusade both sides eventually worked together to petition the government for the scientific success of the project. Indeed, rather than science serving the state, science employed the state to accomplish its goals. Both sides expected to benefit from their mutual cooperation and the tradition of "big" or state-sponsored science was reinforced.

HERSCHEL'S INDUCTIVE THEORY

In addition to these attempts to reform the institutions of science, several prominent natural philosophers sought to reform

---

sciences in Britain, arguing that the "promotion of geomagnetic research thus went hand in hand with advocacy of institutional change." Such a revival he attributes to the "skills and career ambitions of three groups—mathematical practitioners, the Cambridge network, and scientific servicemen," the same individuals who pushed for reform of the Royal Society. The possibility of a venture like the Magnetic Crusade only occurred after members of the "Cambridge network" controlled the "major centers of power in the Royal Society and the British Association." Miller, 107, 129, 134.

35. See for example William Ashworth, "The Calculating Eye: Baily, Herschel, Babbage and the Business of Astronomy." *British Journal for the History of Science*, 27 (1994): 410–440; Roy Macleod, "On the Advancement of Science," *Parliament of Science*. (Northwood: Science Reviews Ltd., 1981): 17–42.

its very methods to create a firmer base for all of science. Establishing a common method for the various branches of science was a way to ease tensions about the increasing fragmentation of different fields and to unify scientific endeavors.[36] Richard Yeo argues that Herschel's work "confirmed the centrality of Bacon's ideas to inquiry about the methodology of science" during a period when Bacon's ideas about scientific institutions and coordinated research finally began to be realized.[37] Dismayed by the fragmentation of Newtonian science, Herschel sought to find a universal philosophical grounding for science that could avoid the problems of induction criticized by philosopher David Hume.[38] Herschel presented his views in the *Preliminary Discourse on the Study of Natural Philosophy*. This volume formed the preface to natural philosophy in Dionysius Lardner's encyclopedia of 1830.[39] Lardner later declared to Herschel that its publication would create a greater sensation than any scientific work that had ever appeared.[40] Herschel, in a more candid moment, admitted that he might have spent more time on the work, had he known the effect it would have.[41] The *Preliminary Discourse* represented Herschel's fundamental views on science, including the methods for universal science that he applied for years to come. Marie Boas Hall believes that the *Preliminary Discourse* expressed Herschel's "proper role in the scientific world." It was "at once a plea for the increased study of the natural sciences... and an analysis of the proper method of pursuing them."[42] Joseph Agassi emphasized the Baconian aspect of Herschel's work as "a major reassertion of inductivism at a time when its status as an account of scientific progress was under threat."[43] For Agassi, Herschel's *Preliminary Discourse* was "a modernized version of Bacon's *Novum*

36. Richard Yeo, "Scientific Method and the Image of Science, 1831–1890," *The Parliament of Science*, (Northwood: Science Reviews Ltd., 1981), 67, 70.
37. Richard Yeo, "An Idol of the Market-Place: Baconianism in Nineteenth Century Britain." *History of Science*, 23 (1985), 251, 267.
38. In the eighteenth century Hume had pointed out that while separate events might seem conjoined, there was no necessary logical connection. "Cause" and "effect" were merely terms and any relationship between them was human assumption. "When we say, therefore, that one object is connected with another, we mean only that they have acquired a connection in our thought and gave rise to this inference by which they become proofs of each other's existence." David Hume, *An Inquiry Concerning Human Understanding*. (New York: Bobbs-Merrill, 1955), 86.
39. Lardner to Herschel, January 19, 1830. (RS HS 11.115).
40. Lardner to Herschel, October 2, 1830. (RS HS 11.119).
41. Herschel to Hall, June 9, 1831. (RS HS 21.82).
42. Marie Boas-Hall, "The 'Distinguished Man of Science.'" *John Herschel 1792–1871: A Bicentennial Commemoration*. (London: Royal Society, 1992), 120.
43. Quoted in Yeo, 1985, 268.

Organon—the same thesis with more illustration, and perhaps with some additional new ideas."[44]

Herschel tried to provide a method for solving the problems of induction, fragmentation and organization. He believed that the "only ultimate source of our knowledge of nature and its laws [was] *experience*."[45] Herschel was a great admirer of Francis Bacon and his method, calling the Baconian philosophy "the grand and only chain for the linking together of physical truths, and the eventual key to every discovery and every application."[46] However, he fully recognized the drawbacks of a purely empirical approach to science. Philosophically, empiricism could not provide the necessity or the general laws which science required. Simply collecting a series of data about phenomenon did not necessarily lead to any natural truth. Inductive logic suffered from its failure to provide certainty of continuity. Nothing guaranteed that a series of similar natural events would continue. Pure empiricism was looked down upon even in the early nineteenth century.[47] No one could claim to base a scientific principle on empirical evidence alone; as a result, some scientists lamented a decline in interest for empirical science.[48]

Herschel appreciated these criticisms of induction and realized that until a way could be found to make inductive conclusions universal, science could not rely upon empiricism.[49] He approached the problem of finding a logical basis for empirical laws. The weakness of induction was its inability to describe generally a phenomena in all cases because of its own limitation to particular cases. Empirically formulated laws could not describe anything beyond the limits of the data from which they were derived. Only deductive, mathematical laws could claim any certainty or necessity, yet Herschel did not trust deduction as a method of science. Deductive philosophers assumed "abstract principles having no foundation but in their own imaginations...with nothing corresponding to them in nature, from which, as from mathematical definitions, postulates and axioms, they imagined that all phenomena could be derived, all the laws of nature deduced."[50]

Herschel recognized that inductive conclusions could not express any natural truth unless they embraced a series of cases that included

---

44. Agassi, 21.
45. John Herschel, *Preliminary Discourse on the Study of Natural Philosophy* (New York: Johnson Reprint Corporation, 1966), 76. Hereafter cited as Herschel, 1830.
46. Ibid., 114.
47. Cannon, 1978, 24.
48. Edward Sabine, "Report on the Variation of the Magnetic Intensity." *BAAS Report* 7 (1837), 66.
49. Herschel, 1830, 167.
50. Ibid., 105.

all of the possible variations.[51] His solution to the problem of induction was universalization. Induction could never lead to a general natural law, unless "particular cases are offered to our observation in such numbers at once as to make the induction of their law a matter of ocular inspection." He used the example of a projectile traveling in a parabolic path. Knowing only a few points through which the projectile passes would not do for making a general law of projectile motion. But by witnessing the entire range of motion, the viewer was able to derive the law that governed its path. "The parabolic form...is a *collective instance* of the velocities and directions of the motions of all the particles which compose it *seen at once,* and which thus leads us, without trouble, to recognize the law of the motion of a projectile."[52] The path of the projectile took it through an infinite number of points; the more of the path that was witnessed, the more readily the applicable law could be found. "As facts multiplied, leading phenomena became prominent, laws began to emerge, and generalizations to commence."[53] That generalized law, describing the individual points of the projectile's flight, could then be expanded to a theory, which covered all the particular points.[54]

From this example, Herschel presented his method for induction. A single observation or single set of observations could not guarantee certainty, but the larger the number of observations of a single phenomenon one collected, the closer and closer one approached a true theory. Every individual case in the series of observations would find its way into the final conclusion.[55] While it might be impossible to ever witness all cases, as one approached the point of witnessing a large number of similar events, one also approached the point of certainty. Herschel believed that there was continuity in nature, and it impressed upon the observer a connection between events.[56] While scientists did not observe continuous phenomena, they could "connect the dots," and see "an ideal outline which we pursue even beyond their limits...thus we arrive at an inductive formula; a general, perhaps universal proposition."[57] "The average of a great many observations... brings us nearer to the truth than any single observation can be relied on as doing."[58] By universalizing the observations across time and place, the theory that was derived could be legitimized.

---

51. Ibid., 177–178.
52. Ibid., 185.
53. Ibid., 116.
54. Ibid., 200.
55. Ibid., 164.
56. John Herschel, *A Treatise on Astronomy* (London: Longman, 1833), 232.
57. Quoted in Thomas Hankins, "A 'Large and Graceful Sinuosity': John Herschel's Graphical Method." *Isis,* 97:4 (December 2006), 621.
58. Herschel, 1830, 215.

Utilizing this analysis, Herschel derived a method for organizing science. He was influenced to a degree by the work of Alexander von Humboldt, whose approach to induction helped shape science in this period. Humboldt sought general theories of science that he did not believe could be based upon local observations.[59] Herschel too held that the achievement of knowledge was too great for a single effort. Rather, stations must be established in order to provide continuous observations in many places [60] The first step of his method was the accumulation of a great number of facts by data collectors. Herschel believed that any well-informed person possessed the power to add something essential to the general stock of knowledge, if he only observed regularly and methodically.[61] The job of data collection was to be democratic and available to almost anyone. As Yeo has argued, "the Baconian conception of science as a collective enterprise provided space for part-time, amateur cultivators of science."[62] Indeed, Herschel himself felt that it was a great pleasure, while traveling, "to have some observation to make to bestow an interest on the many otherwise uninteresting places one passes through."[63] However, Herschel's method contained a division of labor: while fact gathering was a democratic job, the derivation of theory was left to a trained elite.[64] Theory required "a degree of knowledge of mathematics and geometry altogether unattainable by the generality of mankind."[65] While data collection was an open field, Herschel did not accept the more populist Baconianism of Macaulay, whose rhetoric "reinforce[d] popular prejudices about the superiority of common sense over theory and method."[66] For Herschel, "the highest and far the most important pursuits of Science [were] those which are directed towards the improvement of its theories," and theoreticians provided the crucial last link in the chain of inductive science.[67] Thus the collected facts were distilled into theory by a higher level of scientists.[68]

---

59. Cannon, 1978, 80.
60. Herschel, 1830, 175.
61. Ibid., 133.
62. Yeo, 1981, 67.
63. Herschel found his barometer a great comfort at such times. Herschel to Quetelet, December 19, 1831. (APS HS #11:1).
64. Babbage had suggested that a "division of labor was as applicable to science as industry." Quoted in Richard Yeo, "Scientific Method and the Rhetoric of Science in Britain: 1830–1917," *Politics and Rhetoric of Scientific Method.* (Dordrecht: Reidel Publishing, 1986), 266.
65. Herschel, 1830, 25.
66. Yeo, 1985, 272.
67. Herschel to Brisbane, January 2, 1842. (RS HS 22.107).
68. In the Magnetic Crusade, "scientific servicemen led the way in the collection of data, the mathematical practitioners and Cambridge mathematicians contributed to its appraisal and engaged in experimental and theoretical work concerning the 'laws' of geomagnetism." Miller, 127.

The legitimacy of these theories, however, depended not on the experience of one man (or of one generation) but the accumulated experience of all mankind.[69] Inductive conclusions could not claim certainty unless they approached a universal representation of experience, just as multiple points filled in the path of the parabola. This led Herschel to his definition of science: "the knowledge of many, orderly and methodically digested and arranged, so as to be become attainable by one."[70] Herschel carried his universal belief even further, looking for the eventual unification of all fields of science as the final legitimization of science. "There is scarcely any natural phenomena which can be fully and completely explained in all its circumstances, without a union of several, perhaps of all, the sciences."[71] Herschel shared this pansophism with Bacon.

Through his idea of induction Herschel tried to legitimize science through its universalization, demonstrating the truth of a theory by its universal applicability. He did not present necessary ideas as a way to tie together his philosophy. Herschel sought general laws in overlapping fields, rather than analogies between the fields.[72] Herschel relied upon hypotheses to emerge from observation and the instinct to generalize from experiences.[73] He seemed to have received this idea from his father, the astronomer William Herschel:

> it was a saying often in my father's mouth *"hypothesis fingo"* in reference to Newton's *"hypothesis non fingo"* and certainly it is this facility of framing hypothesis if accompanied with an equal facility of abandoning them which is the happiest structure of mind for theoretical speculation.[74]

Herschel held that a collection of data could suggest a physical law from inspection.[75] But at what point did one have the critical mass of observations that allowed you to connect the dots in the parabola and see the overall form of the law? While basing his method on observation, Herschel had no trouble accepting the need for hypothesizing and theorizing from the data. Herschel tried various ways to explain this process. In 1834, he suggested that a sort of divine inspiration reveals laws.[76] In 1840, however, he gave a more creative explanation to Airy that suggested a more human element. The reduction of data to laws

---

69. Herschel, 1830, 76.
70. Ibid., 18.
71. Ibid., 174.
72. Ibid., 61, 95.
73. John Wettersten, "William Whewell: Problems of Induction vs Problems of Rationality." *British Journal for the Philosophy of Science*, 45 (1994), 730.
74. Herschel to Whewell, August 20, 1837. (RS HS 21.228).
75. Herschel to S. Howard, June 16, 1839. (RS HS 22.16).
76. Herschel, 1834.

requires that the observer "think intensely on his subjects and to de-
duce laws from his observations— he must therefore concentrate his
abilities on them and bring the whole force of his mind *as an inventor*
to bear on them."[77] To a degree, Herschel seemed content to ignore the
source of a theory, so long as it applied to the facts and explained the
phenomena.[78] Neither did he appeal to a divine source of innate ideas.
In fact, he saw the scientist as more detached from nature in the role of
a "privileged spectator, humbly but diligently seeking to unravel some
of the lowest of her mysteries, and [to] catch thereby a glimpse, however dim and distant, of the designs of her glorious Author."[79] Another
essential part of his scientific method was the publication of observations to allow theorists of all nations to work.[80]

Herschel's *Preliminary Discourse* became an influential work among
scientists in Britain as well as across the ocean in the United States,
where a reprinted edition enjoyed wide circulation by 1832.[81] By the
time of the Magnetic Crusade, Herschel had modified his earlier belief
that science was better done by individuals and instead enlisted state
aid in order to carry out his plans for legitimizing a geomagnetic theory. Perhaps by substituting the corporate authority of state-sponsored
science for divine ideas, Herschel could legitimize inductive conclusions without recourse to necessary general laws. Herschel was seeking to reaffirm inductive science and to find a way to unify the
increasingly diverse fields of science through a single, universal inductive method.

Herschel had to deal with the practical problem of how to obtain the
network of observations that he required for his theory of universal
science. Like many scientists of his age, Herschel became interested in
large-scale, global observations in the physical sciences. Unlike
Humboldt or Sabine, though, Herschel's interest fit into an overall system of inductive universalism that he had defined years before. In
1830, he wrote that "no natural phenomenon c[ould] be adequately
studied *in itself alone*, but, to be understood, must be considered *as it
stands connected with all nature.*"[82] Almost ten years after the publication of *Preliminary Discourse*, Herschel turned away from mathematics and even astronomy as his interest grew in the geophysical

---

77. Herschel to Airy, July 6, 1840. (UTX 1054:11). Emphasis in original.
78. Richard Yeo, *Defining Science* (Cambridge: Cambridge University Press,
    1993), 96.
79. [Herschel], "Whewell on Inductive Sciences." *Quarterly Review* 66 (1840),
    179. Cannon suggests that Herschel implied that the existence of a ratio-
    nal world was evidence of a benevolent God. Walter F. Cannon, "John
    Herschel and the Idea of Science." *Journal of the History of Ideas*, 22:2
    (April–June 1961), 226.
80. Herschel to Sabine, January 15, 1841. (PRO BJ3/27/38–9).
81. Bowditch to Herschel, June 6, 1832. (APS HS #1:5).
82. Herschel, 1830, 259. Emphasis in original.

sciences: tides, meteorology and geomagnetism. These fields provided
the ideal tests for his system, as they required large numbers of observations from all over the world in order to come up with a global theory that fit the phenomena. It was this goal that led him to propose
adding colonial observatories to the plan of the crusade in 1838.

Herschel was certainly drawing upon a "Humboldtian" observational tradition in his plans for fixed stations, and the influence of
Humboldt upon Herschel cannot be denied. Steven Ruskin argues that
Herschel followed the pattern of "Humboldtian traveling" during his
early days, and that his trip to the Cape of Good Hope can be seen in
this light as well.[83] Herschel's admiration of Humboldt is evident in
his later review of Humboldt's *Kosmos*: "Science has produced no man
of more rich and varied attainments, more versatile in genius, more
indefatigable in application to all kinds of learning, more energetic in
action, or more ardent in inquiry."[84] Yet Humboldt's role in the
Magnetic Crusade must be seen as one of inspiration rather than organization.[85] While Humboldt was seen as the background stimulation
for the crusade, he was not its prime mover in 1838–1839. Indeed,
Humboldt was not even the first to suggest using British colonies as
observing stations (Arago had done that in 1834). Humboldt's effort in
1836 failed to produce the results Herschel's did three years later.[86]

The difference lay in Herschel's ability to combine his own sociopolitical influence with his belief in a system of universal induction
which required large numbers of global observations over a long period
of time in order to work out a universal theory that applied to all observations by eliminating any temporary variations. Herschel's valuable dual position as an "aristocratic" scientist strategically placed
him to deal with elements of both science and state. This position allowed him to push his own ideas about science and to shape the crusade into a form that would meet the needs of his universal inductive
approach through the lobby.

---

83. Ruskin, 12–36.
84. Herschel, "Humboldt's Kosmos." *Edinburgh Review*, 87 (January 1848),
    91. As usual, the review was unsigned but Herschel's correspondence
    makes it clear that he was the author. Empson to Herschel, October 9,
    1847. (RS HS 7.47); Empson to Herschel, December 30, 1847. (RS HS 7.56).
85. John Cawood, "Comments," *Human Implications of Scientific Advance*
    (Edinburgh: Edinburgh University Press, 1978), 146.
86. Malin and Barraclough. "Humboldt and the Earth's Magnetic Field."
    Quarterly Review of the Royal Astronomical Society, 32:3 (1991), 284.
    Based on the evidence of Humboldt's letter to the Duke of Sussex, Malin
    and Barraclough ascribe credit for the resulting colonial stations largely to
    Humboldt without reference to Herschel or Sabine, declaring merely that
    the "famous letter of 1836...quickly bore fruit." Yet to skip from 1836 to
    1839 and attribute the results of half of the crusade to Humboldt's single
    missive ignores a great deal of the story.

Herschel's role in the crusade was key in bringing about a system of stations where Humboldt had failed before him. And while Sabine had called for observations to be made in British colonial possessions in his 1835 plan during the first lobby, he had not intended to establish permanent observatories, merely to use British colonies as convenient stopping points to make observations. Herschel's plan required continuous observations over a long period of time, and thus the existence of permanent structures for observatories. This plan would necessarily involve science with the imperial apparatus, forcing scientists to negotiate for the resources of empire to carry out their observations. The imperial center would coordinate the efforts, but the periphery would generate the raw material. Data processed in Britain flowed in from across the globe, making the observatories a vital part of the system, not merely auxiliary units. New relationships between science and state, between center and periphery, were forming.

CRUSADE

Given his importance in the upcoming lobby, it must not be assumed that Herschel's role was just that of a "celebrity spokesman" for the crusade. Herschel is sometimes portrayed as the public face who represented the lobby while others were doing the real organizational work. Morrell and Thackray see Herschel as the "irreproachable spokesman" for the interests behind the crusade, but call Sabine and Lloyd the "leading" figures in the lobby.[87] John Cawood identifies Sabine as the true "fanatic" motive force behind the lobby, and while recognizing Herschel's "scientific interest in the geomagnetic project" he holds that Herschel's "true value to the magnetic lobby [was] in his prestige and position."[88] Cawood is correct that Herschel's position in society did give him (and the lobby) new access to government officials, but to ignore Herschel's scientific contributions to the crusade is to misrepresent his role.

Herschel became the primary reason that the crusade expanded to include fixed stations for observations around the world in addition to Sabine's proposed Antarctic naval expedition. The proposal to establish the stations changed the very nature of the crusade, making it more than just another temporary magnetic survey. It also tied the crusade more closely to the workings of the British imperial system than an expedition alone could have done. Eventually the crusade was internationalized and provided the basis for continuous physical observations

---

87. Morrell and Thackray, *Gentlemen of Science* (Oxford: Clarendon Press, 1981), 369, 359.
88. John Cawood, "The Magnetic Crusade: Science and Politics in Early Victorian England." *Isis* 70:4 (December 1979), 507.

around the world. All of these elements of the final crusade can be traced back to Herschel's involvement. It will be my goal in this section to demonstrate Herschel's influential role in creating this system of stations and therefore the scientific part he played in the establishment of the crusade.

In their account of the crusade in *Gentlemen of Science*, Morrell and Thackray argue that the plan for the crusade had included stations in addition to an expedition ever since Sabine and Ross had first come up with the idea during their 1818 voyage to find the Northwest Passage. They use Sabine's later (1840) account of the early days of the crusade as evidence for this claim:

> Round about 1818, Sabine also began to nourish the idea of using military officers as scientific auxiliaries in observatories. For almost twenty years, then, Sabine had in his mind the seeds of the two chief aspects of the 1838 lobby: [1] an Antarctic voyage devoted primarily to magnetic research, and [2] fixed observatories manned mainly by army officers.[89]

However, the contemporary evidence does not support either the contention that Sabine had decided on an Antarctic expedition as early as 1818 or that he expanded that idea to include army officers at fixed stations before the time of the lobby for the crusade. Sabine's interest in an Antarctic voyage does not seem to have been piqued until after the discovery of the northern hemisphere "poles" in the 1820s. Indeed, the correspondence of Sabine and Hansteen in that period had focused primarily on the need for equatorial observations, the "*solum desiderium*" according to Hansteen in 1828.[90] Not until the 1835 effort did Sabine write to Hansteen (about his address to the British Association):

> I took occasion to point out the importance of completing our knowledge of the Phenomena of the Earths Magnetism: and that as the present period is distinguished by your determination of the place of the Siberian Pole [1829], and by that of Captain James Ross of the place of the N. American Pole [1831], it shall not be supposed to pass away without a corresponding determination of the positions of the two Poles in the Southern Hemisphere.[91]

Additionally, the idea of using army officers at the stations seems to have been a later adaptation to the reluctance of the admiralty to go along with the fixed observatories. Herschel had initially questioned the employment of officers, arguing that permanent and systematic observations could "hardly be expected from Military officers having other and important duties and liable to frequent removal."[92] It was

---

89. Morrell and Thackray, 354.
90. Hansteen to Sabine, April 20, 1828. (PRO BJ3/3/19).
91. Sabine to Hansteen, August 19, 1835. (PRO BJ3/8/2).
92. Herschel to Minto, October 20, 1838. (UTX 1054:263).

not until they drew up the plan for the observatories in November 1838 that Sabine and Lloyd mentioned specifically that the observers were to be officers. Even then, only the observers at St. Helena and Canada were to be from the army. The Tasmanian observatory was always to have been staffed by the navy.[93] It was not until the spring of 1839, with the crusade approved, that Sabine began to look into the official arrangements needed for the use of artillery officers.[94] It seems that if it had truly been Sabine's intention to use officers for observers all along that he might have made some effort of determining, prior to this point, whether or not such an arrangement might be acceptable to the artillery.

As for the inclusion of the stations in the plan, according to Sabine's recollection, it was Lloyd who was to press the idea on Herschel as part of their appeal to him to join the lobby in August 1838. Sabine visited Lloyd at Dublin in the summer of 1838 to discuss observing stations "appropriate to each of the principal magnetic centres."[95] But Lloyd's letter to Herschel of August 4, 1838 (cited by Morrell and Thackray as Lloyd's push for observatories), rather emphasized Herschel's "strong opinion of the importance of a South Polar expedition," and mentioned the establishment of only "one or two magnetical observatories in British India."[96] This last point may have been a reference to Jervis's attempt at conducting observations in the East India Company holdings, which Lloyd had heard about earlier that year. Judging from Lloyd's letter, the observatories were a secondary part of the plan, not the extensive system that Herschel was nurturing.

It is my argument that Herschel's contributions to the Magnetic Crusade were integral and essential. His philosophical beliefs concerning inductive science led him to add his own distinct and crucial additions to the plan of the crusade, especially in the form of fixed observatories. He made it possible for scientific goals to be accomplished through the resources and structure of the British imperial system. Like the British Empire, Herschel's science was cosmopolitan and universalist. While his political and social connections were invaluable for the lobbying effort, he was not just its "celebrity spokesman." Nor was he merely drawn into an existing plan that had been maturing in Sabine's mind for twenty years. Even Cannon is mistaken when he states that Herschel was "lured" into the crusade.[97] As will be seen, Herschel had in fact been harboring similar plans for many years.

---

93. Lloyd to Herschel, November 13, 1838. (RS HS 11.266).
94. Sabine to Herschel, March 15, 1839. (RS HS 15.28).
95. Sabine to Phillips, July 4, 1838. Reprinted in *Gentlemen of Science: Early Correspondence of the BAAS*, Jack Morrell and Arnold Thackray, editors (London: Royal Historical Society, 1984), 263.
96. Lloyd to Herschel, August 4, 1838. (RS HS 11.265).
97. Walter Cannon, "John Herschel and the Idea of Science." *Journal of the History of Ideas*, 22:2 (April–June 1961), 219.

It is clear that Herschel came to the idea of using stations for worldwide observations on his own, without any prodding from Sabine or
Lloyd. After the publication of the *Preliminary Discourse*, he developed his inductive ideas during the 1830s. Herschel's interest in global
observations can be traced to his particular view of universal induction, which held that while general propositions were by their nature
only probable truths when applied beyond the range of the instances
from which they were derived, the more frequently they were tested in
other instances, the more their probability approached certainty.[98] For
terrestrial magnetism, Herschel defended the need for stations to acquire continuous observations because the variations in the magnetic
field could only be detected over many years. He also held that the
data must be collected simultaneously at the various stations, in order
to treat the variations globally. He believed that with enough data from
particular points, a full theory of geomagnetism could be worked out
for the whole world.

> The secular variations of the constant elements in a general theory of
> terr[estrial] Magnetism is one of indispensable importance and *can* only
> be obtained by observations of extreme precision carried on for many
> years Were they known for every point on the globe we might of course
> by their aid bring all ancient and modern obs[ervations] to one epoch.[99]

Herschel had been active in plans to set up extensive series of geophysical observations for some years before the crusade. One of his
first actions upon arriving at the Cape was to make a series of meteorological observations.[100] In 1835, he had written to William Henry
Smythe, recommending a new plan for hourly worldwide meteorological observations on four predetermined days each year all over the
world. He and Thomas Maclear (the observer at the Capetown observatory) were planning to participate from the Cape, and he hoped it
would be undertaken in India, New South Wales, Mauritius, and in
many other locations.[101] In 1837, he gained the cooperation of Adolph
Quetelet (the Belgian statistician and meteorologist) for a series of simultaneous meteorological observations between Europe and the
Cape.[102] Herschel was also in correspondence with John Webster in the
United States (to whom he recommended a series of observations for
1835 and 1836)[103] as well as the Albany Institute, from which he received a set of meteorological observations for the state of New York.[104]
(At the time, Herschel was "delighted to find...in what a liberal and

98. Herschel, 1834.
99. Herschel to Lloyd, November 5, 1838. (RS HS 21.267).
100. Herschel to Paine, August 26, 1836. (UTX 1054:291).
101. Herschel to Smythe, May 10, 1835. (RS HS 21.175).
102. Herschel to Quetelet, June 15, 1837. (APS HS #11:1).
103. Webster to Herschel, February 16, 1837. (RS HS 18.128).
104. Herschel to Albany Institute, September 30, 1836. (APS B:H435p).

indeed splendid manner the American government [wa]s pursuing scientific objects."[105])

While at the Cape, Herschel collected meteorological observations from around the world. Between 1835 and 1839 Herschel received reports from India, Geneva, Guyana, Mauritius, Turin, Albany, Port Arthur, Boston, Tasmania, Bermuda and Brussels. In 1836 he became an honorary member of the British Meteorological Society. He also intended the system of observations to continue after his departure, publishing a pamphlet on making meteorological observations in southern Africa and the South Seas shortly before he left in 1838. Such observations were to be taken at the Cape, Ascension Island, Mauritius, Tristan d'Acuna, Madagascar and Mozambique.[106] Herschel's involvement in meteorological studies on an international scale prepared him for similar efforts on behalf of geomagnetism later.[107]

Herschel also saw the tides as a field worthy of study, especially as his friend William Whewell was making a study of them for a book. Whewell (1794–1866) had met Herschel at Cambridge and the two remained friends. Whewell remained at the university, serving as master of Trinity College. Like Herschel, he was concerned with legitimizing inductive science.[108] Whewell was best known for his *History of the Inductive Sciences* (1837) and his *Philosophy of the Inductive Sciences* (1840).[109] Herschel assisted Whewell by gathering tidal observations both at the Cape and from his brother-in-law in Canton.[110] Ironically,

---

105. Herschel to Baily, April 7, 1837. (RS HS 25.8.10).

106. John Herschel, *Instructions for Making and Registering Meteorological Observations in South Africa.* (London: Bradbury and Evans, 1838), 17.

107. Giuliano Pancaldi, "Scientific Internationalism and the British Association," *The Parliament of Science* (Northwood: Science Reviews Ltd., 1981), 153.

108. Unlike Herschel though, Whewell turned to innate ideas as a foundation of inductive science, showing a clear Kantian influence. He rejected the "Baconian notion of an accessible mechanical set of rules which produced reliable knowledge." Yeo, 1985, 276. "My argument is all in a single sentence," claimed Whewell. "You *must* adopt such a view of the nature of scientific truth as makes universal and necessary propositions possible; for it appears that there are such, not only in arithmetic and geometry, but in mechanics, physics and other things. I know no solution of this difficulty except by assuming *a priori* grounds." Whewell to Quetelet, June 28, 1839. (WC). However, Whewell was able to combine ideal and empirical methods in his system. For Whewell, induction was the process of discovering truth not through inference from facts, but from the imposition of a necessary idea. For Whewell and induction see Laura Snyder, "It's *All* Necessarily So: William Whewell on Scientific Truth." *Studies in History and Philosophy of Science*, 25 (1994).

109. DSB.

110. Herschel to Stewart, November 25, 1835. Reprinted in *Herschel at the Cape*, Evans, David S., editor. (Austin: University of Texas Press, 1969), 198; Herschel to Whewell, July 22, 1837. (RS HS 21.224).

only in the field of geomagnetism did Herschel fail to make any observations at the Cape. Writing to his aunt Caroline in 1836, he explained that while he had heard of Gauss's new method for making geomagnetic observations, he lacked the necessary equipment to carry them out in Africa.[111] Despite this lacuna, Hankins argues that Herschel returned to England "experienced in organizing the kind of measurements required for mapping terrestrial magnetism even though he had done no magnetic work himself."[112]

Herschel saw the value in setting up "physical observatories" around the world for geophysical observations. In 1835, he wrote extensively to Captain Francis Beaufort, the royal hydrographer, about the nature of such physical observatories. Complaining about the proliferation of colonial astronomical observatories, Herschel maintained that small, local observatories could contribute little to astronomy. Rather, he believed that such establishments should be converted to physical observatories to make observations concerning magnetic intensity and direction, meteorology and the tides. Such observatories could also serve as centers to propagate accurate standards of weights and measures in the colonies[113] (and indeed Herschel mentioned this when writing to David Forbes in 1836[114]). These goals were among those he had laid out in *Preliminary Discourse*.

Although he was unable to make geomagnetic observations while in Africa, Herschel did busy himself helping to set up observatories around the world to make geomagnetic, in addition to tidal and meteorological, observations. In 1836 he played a role in the establishment of a new observatory in Bombay. In 1837 Herschel advised George Gipps to set up a physical observatory in Australia for tidal, magnetic and meteorological observations, rather than an astronomical one.[115] Similarly in mid-1838 he advised Captain Beaufort on establishing a physical observatory on Mauritius.[116] It is worth noting that all of these efforts came before Herschel's involvement with the Magnetic Crusade.

111. Herschel to Caroline, March 8, 1836. Reprinted in Evans, 214. Here Herschel apparently forgot an earlier promise to Quetelet. Writing in 1831, Herschel had expressed interest in Quetelet's new magnetic observations, and vowed to "make a case of magnets a principal feature in my list of apparatus" should he travel again. Herschel to Quetelet, December 19, 1831. (APS HS #11:1).
112. Hankins, 2006, 622.
113. Herschel to Beaufort, October 11, 1835. (RS HS 21.188).
114. Herschel to Forbes, November 15, 1836. (SAUL msdep7, Incoming Letters 1836 #60).
115. Herschel to Gipps, December 27, 1837. (RS HS 21.235).
116. Herschel to Beaufort, June 29, 1838. (RS HS 21.253); Herschel to Beaufort, July 22, 1838. (RS HS 19.74).

Herschel's part in setting up observatories to provide the raw data for his universal inductions may have caused him to change his mind about the individual nature of scientific activity. In 1831, he had argued that state funding for science was more than adequate.[117] Now realizing the value of the state in the task of setting up colonial stations, Herschel worked to bring more state support to science. While he opposed "the *frittering away of large sums* on no well digested & concerted plan," Herschel did declare himself in favor of "detecting great & worthy objects, stating them in their broadest form and then going the whole hog at once in the question of expense."[118] State assistance not only provided the necessary sites for observing stations, the state could also provide crucial coordination for simultaneous observing. Herschel saw the state as the patron of science, but did not believe that the state was necessarily required to back all scientific ventures. For this reason he did not support every proposal for an observatory that came to him.[119] Herschel's view toward the state followed a courtier model in which the state served as the support for scientific operations in return for scientific benefits and prestige, a fitting view for the son of the British court's premier astronomer. The state could provide backing and logistic support for science, but Herschel did not see the state as the legitimizer of science as did Sabine.

With the beginning of the lobby and a chance at state support for his plans, Herschel reached one of the critical points in his scientific career. Until then, he had been helping to set up observatories around the world in a private capacity by advising and writing letters of support. But his overall plan of a more extensive global system of stations for the purpose of observing seemed out of reach. In 1835 when he had laid out his plans for physical observatories to Captain Beaufort, he had ended his letter by commenting that "perhaps all of this is dreaming."[120] Writing to Forbes in 1836, he had expressed his desire to see a new class of physical observatories established, but regretted that he did not have time to commit such a project.[121] His work in South Africa kept him too busy to become more involved in physical observations, but on his return to England in the spring of 1838, he was finally able to devote his time to it.

Thus Herschel already harbored a desire to create a system of global observatories in the British colonies by the summer of 1838, before the lobby for the crusade had even begun. While in Africa, he came to

---

117. Herschel to Hussey, August 2/3, 1831. (RS HS 25.2.11).
118. Herschel to Airy, November 15, 1839. (UTX H/L:10).
119. Herschel to Wheatstone, June 17, 1842. (RS HS 18.149). But he did believe that the state had a role in supporting scientists so that they could work without having to teach. Herschel to Beaufort, October 11, 1835. (RS HS 21.188).
120. Ibid.
121. Herschel to Forbes, November 15, 1836. (RS HS 21.213).

accept the utility of observing stations in the British colonies. Eventually he embraced the involvement of the state in science as necessary as he looked to the British colonial world to supply sites for observatories. Accepting that his individual efforts were insufficient to set up enough stations to furnish the observations that could provide inductive legitimacy, he decided to go directly to the British government. In the absence of the now dissolved Board of Longitude, he appealed through another institution that had an interest in geomagnetism, the admiralty.

In an important letter of June 1838, shortly after his return from Africa (but still before the Magnetic Crusade lobby), Herschel laid out his proposal to his friend Captain Francis Beaufort. Citing the recent success of the European geomagnetic stations, he proposed the establishment of a similar system of observations "over the whole surface of the globe, and especially of establishing permanent magnetic stations at the Cape, in India, Australia and other points within the range of British superintendence." He stressed the need for stations in addition to new voyages of discovery which had become more common in recent years. He particularly concentrated on the field of geomagnetism; he argued for the importance of a knowledge of terrestrial magnetism and for observations in the southern hemisphere, where less was known. These observations corresponded to regions of British colonial expansion in Africa and Australia as well as in India.[122]

This letter is important because it shows that Herschel had already reached the conclusion that a global system of observation points was necessary to provide data, even before Sabine and Lloyd began to lobby for the same objective. He even appealed to Humboldt to help him in his private efforts to bring the British state on board. Herschel hoped that Humboldt would back his plan for a series of colonial observatories at the Cape, India, Australia and Mauritius ("in short as many stations as possible in the English colonial possessions") in correspondence with those in Europe. [123] While Herschel felt he had every reason to hope that his suggestion might be adopted, his private appeal came to nothing. But Herschel's failure to convince the admiralty to go along with his plan in the summer of 1838 provides an indication as to why he joined the lobby for the Magnetic Crusade in the fall.

122. Herschel to Beaufort, June 29, 1838. (RS HS 21.253).
123. Herschel to Humboldt, July 31, 1838. (RS HS 21.255).

Chapter Three

Worthy a Great National Undertaking

The men involved in the magnetic lobby were primarily connected with the new British Association for the Advancement of Science as well as with the established Royal Society. These personalities behind the crusade were often as important as the ideas. Herschel and Sabine were well connected with the ruling land-owning and military classes and knew how to lobby for their project; Herschel was a baronet and prominent member of the gentry class while Sabine was an artillery officer and graduate of the Royal Military Academy. It will be my argument that Herschel's involvement with the crusade in 1838–1839 was both significant and necessary for its success in the form it took. Herschel's contributions to the crusade were important, especially in dealing with the issues of access to influential people and the establishment of the fixed observing stations. Herschel's efforts were critical to the success of the 1838 lobby. Compared to the failures of the 1830s, the proposal of 1838 was well received and enthusiastically adopted by the state. By tracing the course of the lobby for the crusade, it is possible to see the importance of both scientific ideas and political connections in the construction of this important scientific project.

▓ THE LOBBY

The beginning of the lobby can be found in Sabine and Lloyd's joint letters of August 4, 1838. The target of its first round of rhetoric was not the British government or the British Association, but Herschel himself. Knowing that Herschel was reluctant to become involved in such a large undertaking, Sabine and Lloyd each addressed a letter to him appealing for help. As both letters were written on August 4 and both were from Dublin, we can accept Sabine's later claim that there was some definite collusion between himself and Lloyd to gain Herschel's aid.[1] The similarities in the letters leaves almost no doubt that Herschel had become the target of a well-organized plan to recruit him for the lobby.

Lloyd wrote that three years had passed since the last attempt at a magnetic project. Both Lloyd and Sabine now argued that the time had come for another attempt. The important discoveries that had resulted from the system of simultaneous observation under the direction of Gauss and the new results obtained by Sabine in the Arctic had given a fresh impulse to magnetic investigation.[2] Sabine held that the time had arrived for new magnetic research on a global scale.[3] Lloyd suggested that Britain could make up the difference between British science and that of the continent within the next three years. Both intended that the suggestion should be brought up at the upcoming British Association meeting in Newcastle (over which Herschel was originally meant to preside), and prevailed upon Herschel to present the idea. Lloyd claimed that there was every reason to believe that the government would act if the British Association recommended the project. But both were also convinced that only Herschel could successfully bring about government aid for the crusade.

Sabine and Lloyd knew that they had to convince Herschel to join the lobby, and thus used their best arguments. Despite these attempts, Herschel's first response was to insist that Lloyd was better suited to lead the lobby than he.[4] For a variety of reasons, Herschel was initially hesitant to join the lobby. Having just returned from observing the southern hemisphere at the Cape of Good Hope, he had five years of astronomic observations to reduce for publication and was worried that the lobby would take him away from that work. Indeed later, Herschel noted that the magnetic project had "eaten up a year of my life and thrown me back in all my projects...in some irrecoverably."[5]

---

1. Sabine to Herschel, July 7, 1840. (PRO BJ3/26/136).
2. Lloyd to Herschel, August 4, 1838. (RS HS 11.265).
3. Sabine to Herschel, August 4, 1838. (RS HS 15.18).
4. Herschel to Lloyd, August 6, 1838. (RS HS 21.256).
5. Herschel to Whewell, August 6, 1839. (RS HS 22.24).

Personally, Herschel did not want to become bogged down in official business or be restricted so that he could not pursue his own interests. He turned down both a teaching position at Oxford and a magistracy because of this attitude.[6] Family life (he had a wife and twelve children) also kept him busy.

Nevertheless Herschel did decide to join the lobby for the crusade. Several reasons can be adduced for this decision. First, during his time in Africa Herschel had been in contact with Gauss through his aunt Caroline. It was in a letter to Caroline that Herschel first expressed interest in Gauss's new theory.[7] Although he was unable to make geomagnetic observations at the Cape because he lacked the equipment, Herschel was anxious to learn more about geomagnetism.[8] In June, shortly after Herschel's return to London, Wilhelm Weber—a German scientist who was involved with Gauss and Johann Poggendorf in geomagnetism—had visited him. Weber had come over to organize a system of corresponding magnetic observations on Gauss's plan.[9] This meeting led Herschel to visit Gauss himself during a trip to Hanover the next month.[10] The timing of these visits helped to reinforce this new interest in geomagnetism. Finally, Herschel must have seen the Magnetic Crusade as a chance to continue his efforts toward establishing a global system of observations for the physical sciences that had begun with his appeal to the admiralty in the summer of 1838. His belief in the importance of the stations and his influence in their establishment will be seen in the course of the lobby.

In the initial planning stages of the crusade, several problems emerged. Sabine and Herschel found themselves on opposing sides of a theoretical division. Like Hansteen, Sabine held the four-pole theory and believed that magnetism was a cosmic force. Hansteen's theory influenced Sabine's choice of expeditions because it favored certain locations for the poles. Observations in these places were crucial, because only there did the magnetic poles exist.[11] With exploration of the northern hemisphere complete and magnetic records made, Sabine advocated that the Magnetic Crusade concentrate on the southern

6. Beaufort to Herschel, March 28, 1839. (RS HS 3.363); Herschel to Micklewait, January 13, 1841. (RS HS 22.76).
7. Herschel to Caroline, October 24, 1835. Reprinted in David S. Evans, editor. *Herschel at the Cape.* (Austin: University of Texas Press, 1969), 191.
8. Herschel to Caroline, February 19, 1836. Reprinted in Evans, 214; Herschel to Caroline, March 8, 1836. Reprinted in Evans, 218.
9. Herschel to Lloyd, June 11, 1838. (RS HS 21.248).
10. The Herschel family came from Hanover originally and Herschel's aunt Caroline still lived there. William Herschel had come to England because of his connection to the Hanoverian court of George III. Diary: July 21, 1838; DSB.
11. Sabine, "Report on the Variation of the Magnetic Intensity." *BAAS Report* 7 (1837), 77.

hemisphere, especially in the predicted vicinity of Hansteen's southern poles.[12] Thus he recommended an Antarctic expedition.

Herschel, on the other hand, adopted the view of Gauss.[13] Since Gauss held that there was only a single axis, his theory did not have any privileged points for observation.[14] Due to this belief, Herschel did not feel that there was "any one geographical point to be pushed for in preference to another."[15] Gauss himself had advocated further observations in order to further refine his theory of terrestrial magnetism and so Herschel sided with Sabine to push for the crusade.[16] His interest, however, was more directed toward establishing a series of physical observatories for simultaneous magnetic readings around the world rather than a single push to the Antarctic.

On the practical side, there were also opposing influences at work. Some members of the lobby pushed the nationalist character of the crusade, portraying it as an avenue for British science to catch up to the rest of Europe, a position that appealed to the many scientists who had come to the conclusion that Britain was behind Russia, France and the German states in science. Others saw the crusade as an international, cooperative venture that required British efforts to be combined with those of other European nations. Some members of the lobby pressed the practical applications of geomagnetism: benefits to navigation and commerce, while others dismissed them and emphasized the overall benefits to science instead. Generally speaking, Herschel and Lloyd tended to represent the Gauss/internationalist/observatory side of the lobby while Ross and Sabine represented the Hansteen/nationalist/expedition side. The consequences of this division threatened the success of the lobby in later months.

The lobbying efforts for the Magnetic Crusade were initially tied into the activities of the new British Association for the Advancement of Science. Complaining of the decline of science and Britain's failure to keep up with the continental states, the British Association lobbied the British government on behalf of several scientific enterprises, including the new Ordnance map survey of Britain and the Magnetic Crusade.[17] New advances in geomagnetism seemed to be coming at

---

12. Ibid., 82, 90.
13. John Cawood, "Terrestrial Magnetism and the Development of International Collaboration in the Early Nineteenth Century." *Annals of Science* 34:6 (November 1977), 586. Whewell also agreed with the two-pole theory. William Whewell, *History of the Inductive Sciences* (London: John Parker, 1857), III:54.
14. John Herschel, "Terrestrial Magnetism." *Quarterly Review*, 66 (1840), 283.
15. Ibid., 312.
16. G. D. Garland, "The Contribution of Carl Friedrich Gauss to Geomagnetism." *Historia Mathematica* 6 (1979), 18.
17. Cawood, 1977, 584.

such a fast pace that the completion of the theory seemed only a matter of time. In 1838 at their meeting in Newcastle, the British Association passed a resolution calling upon the British government to support an expedition to the southern hemisphere as a first step in completing the geomagnetic theory.[18] The committee which put together this proposal included Herschel, Sabine, Lloyd, Whewell and George Peacock.[19] By this point Sabine and Lloyd had also joined Herschel's push for the establishment of magnetic observatories in British colonies to accompany the expedition.[20] This committee formed the nucleus of the magnetic lobby, which pushed for the crusade through meetings with government officials, efforts with the admiralty, and eventually on the pages of the *London Times*. The lobby for the Magnetic Crusade used a combination of sometimes contrasting arguments in order to sell the crusade to the government.

The committee's proposal first made it clear that the limits of private scientific research had been reached. The proposal claimed that the British Association had approached the government only because individuals could not accomplish the task.[21] The necessity of employing simultaneous observations in combinations to elicit the values of the magnetic constants required governmental support and coordination.[22] As for the task involved, the association felt that it was in the national interest. They would not ask national support except where the object aimed at was of national importance.[23] Sabine held that the most important service that Britain or any other country could render to geomagnetism would be to fill the void in the southern hemisphere.[24] There was no hope for a general geomagnetic theory until such southern observations were made.[25]

In selling the crusade, the lobby focused both upon scientific and practical goals. Its members claimed that the consequences of the crusade were the general advancement of human knowledge and a universal impulse to other branches of science, leading to a higher standard of

18. John Herschel and Humphrey Lloyd, "Report on the subject of a series of Resolutions adopted by the British Association at their Meeting in August, 1838, at Newcastle." *BAAS Report 9* (1839), 31. Whewell, 1857, III:50.

19. George Peacock (1791–1858) taught math and astronomy at Cambridge. He became dean of Ely Cathedral in 1839. Whewell and Peacock, both at Cambridge, played only a minor role in the lobby and usually supported Herschel.

20. Sabine later claimed that at the time of the Newcastle meeting only he, Lloyd, Herschel and Beaufort had known how extensive the role of the observatories would be. Sabine to Lloyd, December 20, 1839. (RS Te #80).

21. "Report of a Committee." *BAAS Report 10* (1840), xxxvi.

22. Herschel and Lloyd, 1839, 34.

23. Ibid., 38.

24. Sabine, 1837, 84.

25. Herschel and Lloyd, 1839, 35.

physical investigation.[26] But the lobby also stressed practical elements of the crusade, primarily contributions to navigation and discovery.[27] The magnetic impact upon iron vessels ("under any circumstances important to a great maritime power")[28] was one practical field that could be studied by the crusade, and there was always the possibility of new geographic discoveries in the Antarctic.[29]

But even enthusiasts privately admitted that the practical effects on navigation would be limited. Herschel's interest in geomagnetism helped to acquire government support for the crusade, but the magnetic expedition was always about science first. The crusade itself had no immediate practical application in improving navigation. The very fact that the expedition was expected to reach previously unexplored points in order to take unknown magnetic readings seemed to indicate that navigation was already in advance of any aid which geomagnetism could offer. Herschel later admitted that the expedition would pass through regions that "are little accessible, and unlikely to be visited by commercial intercourse or enterprize," and thus would be of little use to navigation or to a maritime and commercial nation.[30] Herschel himself recognized the futility of trying to improve navigation charts with magnetic observations. It had been realized centuries ago that the variations of the compass themselves varied annually, and that only by making constantly newer sets of observations could these variations be accounted for. The variable coefficients of Gauss's equations could only be acquired through observation.[31] One had to go and make observations of the magnetic field before one could calculate what the field should be in that point. These problems could only be overcome by a complete theory of geomagnetism that could predict the secular changes in the magnetic field at points in the future, rather than having to rely upon continued observations at different places to determine those values. Benefits to navigation were emphasized primarily to sell the crusade to the navy.

Beyond these promised practical results, the magnetic lobby also appealed to the vanity and nationalism of the state while at the same time hoping for international cooperation among the various nations interested in geomagnetism. By 1835, French and German magnetic

26. *BAAS Report*, 1840, xxxvii, xxxix; Herschel, 1840, 296.
27. Herschel and Lloyd, 1839, 37.
28. Ibid., 38.
29. Herschel also claimed that "theoretical knowledge has uniformly been followed by a new practice and by the abandonment of ancient methods as comparatively inefficient and uneconomical," and that completing the magnetic theory would surely have an impact upon navigation. Herschel, 1840, 296.
30. Ibid., 306.
31. Ibid., 281.

observatories were more advanced than Britain's.[32] Even Norway and Russia seemed to have contributed more to the study of geomagnetism.[33] This nationalistic rhetoric encouraged the British government to take the lead in the quest for a complete geomagnetic theory that would benefit navigation by allowing variations in the geomagnetic field to be predicted in advance. Britain would only be fulfilling its rightful place among nations. Halley had begun the work 150 years earlier, and Britain now picked up where he had left off.[34] Herschel declared, "no nation was ever so favorably situated for such a purpose, nor so strongly called on as a maritime and commercial country for cooperation in a cause directly connected with nautical objects."[35] This nationalistic appeal seems to have been employed mainly to achieve government support, as the British Association itself hoped that Britain's example would be followed up by other nations, and that the operation would be not merely a British, but a European and American, one.[36] Herschel was also wary of an overly nationalistic tone, preferring to call for an international effort to collect the data.[37]

Here there was an interesting contrast of national and cosmopolitan science, a tension that created something of a split personality in the lobby. The British Association had to rely upon nationalist appeals in order to sell the crusade, but preferred a more international approach to science. In addition, the association was also split along theoretical lines in its advocacy of the crusade. These divisions within the lobby caused internal problems that threatened to upset the entire project. Before their meeting with the government, Herschel had to appeal to the members of the lobby to present a united front lest the appearance of internal division should discourage the government from taking their request seriously.[38]

By October 1838 Ross, already eager for the expedition to be approved, wrote to Sabine that:

> to the South there is indeed a most a most [sic] splendid harvest...[yet] unmapped. In every department of Science and I fear that I should envy any other person, who might be so fortunate to get there before me. Has anything been done by the Committee that was appointed to confer with government on the subject?[39]

32. Cawood, 1977, 585.
33. Herschel, 1840, 295.
34. Sabine, 1837, 85.
35. Herschel, 1840, 294; for Sabine it was "a fitting enterprise of a maritime people" to complete geomagnetic theory. Sabine, 1837, 85.
36. Sabine, 1839, 42.
37. Herschel, 1840, 293.
38. Herschel to Airy, November 15, 1838. (UTX H/1:10); Herschel to Lloyd, November 16, 1838. (RS HS 11.267).
39. Ross to Sabine, October 10, 1838. (PRO BJ3/16/16).

Following the Newcastle meeting, Herschel continued his role in the lobby on several courses. He applied to meet with the Prime Minister, Lord Melbourne, in late August 1838.[40] By late September, Herschel had received a commitment to meet with Melbourne when he returned to London.[41] On October 15, Herschel dined with the prime minister and Queen Victoria, discussing Ross's polar expedition with them.[42] That same week, Ross reported that he had a note from Beaufort at the admiralty that told him enough about the Antarctic Expedition to make him anxious to hear more. "It is evidently favorably received thus far by the Government or they would not trouble themselves to ask any opinions," Ross assured Sabine.[43]

Herschel was preparing to attack the issue on several fronts. He appealed to other allies, to which he assumed the government might turn for an opinion. Herschel predicted that a direct application to the prime minister would lead the government to turn for advice to the two bodies that had the most interest in the topic, the admiralty and the Royal Society ("at least in the good old times they would have done so"). Herschel envisioned the process eventually coming back around to the Royal Society, the unofficial scientific advisors to the government. By preparing his numerous allies in those bodies in advance, Herschel could be fairly confident of a good response even before the initial appeal had been made.[44]

On October 20, he addressed a letter to Lord Minto at the admiralty, pressing the need for permanent magnetic stations in regular correspondence with those in Europe as well as an Antarctic expedition. Herschel tailored his appeal for Minto, and emphasized the practical and nautical possibilities of the crusade. Geomagnetic theory "cannot but prove productive of the most momentous results, especially to a great maritime nation to which a complete knowledge of the laws of terrestrial magnetism can never be the object of small practical consideration." Herschel was also quick to suggest the interesting lines of discovery that such an expedition might enable.[45] By the end of October, things seemed to be going well; Herschel wrote to Whewell that Melbourne was not opposed to the project.[46] Sabine, however, thought that the government would try to cut costs and cautioned that

40. Herschel to Lamb, August 27, 1838. (UTX 1054:257.1).
41. Herschel to Sabine, September 22, 1838. (PRO BJ3/3/3).
42. Herschel to Whewell, September 13, 1838. (RS HS 21.258); Diary: October 15, 1838.
43. Ross to Sabine, October 19, 1838. (PRO BJ3/16/52).
44. Herschel to Whewell, October 22, 1838. (RS HS 21.264).
45. Herschel to Minto, October 20, 1838. (UTX 1054:263).
46. "I should presume that the project will at least meet with no opposition on his part, as his enquiries respecting it were such as to manifest considerable interest in the matter." Herschel to Whewell, October 22, 1838. (RS HS 21.264).

they should keep to the goals that could not be obtained by private means.[47] This important point, that only by state aid could the crusade happen, was central to the lobby.[48]

In October Herschel asked Whewell, Peacock, Lloyd and Sabine to draw up a program for the crusade according to their own ideas.[49] Lloyd drew up a list of duties for the observatories. He hoped that with a little pressure, the instruments might be completed and the observers trained before the sailing of the expedition. Ross would then be able to take them out with him, and to deposit them on his way at the various colonial stations. [50] Sabine contributed a plan for the expedition.[51] Everything seemed to be moving smoothly from their point of view. Ross was also confident of success.[52]

Yet trouble arose from the admiralty.[53] Minto's reply to Herschel, while generally favoring the project, queried the cost of the stations. He pointed out that various surveys cost the country upwards of £36,000 a year, and that the admiralty was already spending £60,000 a year on science. Due to the great increase in expenses in recent years it was necessary to consider economies for any new undertaking. Minto suggested that the existing surveyors in the colonies might instead carry out the observing work of the crusade.[54] This comment seems to have thrown Herschel into a slight panic. In a hasty postscript to Whewell, written just after receiving Minto's letter, Herschel wrote "a note just received from Lord Minto is much in the nature of a wet Blanket on the whole concern." It was clear that support for so grand a project would not be easily extracted from the state. The lobby would have to make an all-out effort to convince the government of the merits of the crusade. Seeking Whewell's full commitment, Herschel queried, "are you fully prepared to declare that the objects proposed are really worthy a great national undertaking involving much expense and to defend the expenditure tooth and nail?"[55]

One can see why Herschel was so discomforted. Minto had completely misunderstood the purpose of the crusade. Instead of the global

---

47. Sabine to Lloyd, October 23, 1838. (RS Te #58).
48. Sabine's Memo, October 24, 1838. (PRO BJ3/15/4–5).
49. Herschel to Lloyd, October 22, 1838. (RS HS 21.263).
50. Lloyd's Memo, October 25, 1838. (PRO BJ3/15/4–5).
51. Sabine's Memo, October 24, 1838. (PRO BJ3/15/6–7).
52. Ross to Sabine, November 5, 1838. (PRO BJ3/16/64).
53. In addition to Minto's reluctance, Sir John Barrow (permanent secretary of the admiralty) was unenthusiastic about the crusade due to his preference for an Arctic expedition. Alfred Friendly, *Beaufort of the Admiralty* (New York: Random House, 1977), 291.
54. Minto to Herschel, October 24, 1838. (UTX 1087:372). The British were undertaking a geographic survey of India at this time. See John Keay, *The Great Arc* (New York: HarperCollins, 2000).
55. Herschel to Whewell, October 28/30, 1838. (RS HS 21.266).

system of observations that Herschel had envisioned, Minto saw it as
no more than another naval survey. Replying to Minto, Herschel made
a spirited defense of the need for the stations, not only because of
their immediate value to the crusade, but also their potential to stimulate further research in other related fields which could be studied
through them.

> Is it hoping too much that the day may not be far distant when Physical
> Science in all its exacter branches shall participate in these advantages
> and when the establishment of "Physical Observatories" in our own and
> distant lands shall give that impulse to many other sciences (as for exam-
> ple Magnetism, Meteorology, etc.) of which they stand so much in need?[56]

Others were not as concerned as Herschel about the course of the
lobby. Ross felt that it was going well, if slowly. He felt that the fixed
observations were a much simpler affair than Herschel seemed to
think, and that they could be easily established at the time of the crusade. "To talk of deferring it to another year is next to madness. There
is abundance of time to do all if the decision be prompt."[57] Robert Fox
could not believe that the government would refuse to send out an expedition for the attainment of an object that he saw as so important in
every point of view.[58] During November 1838 planning for the observatories continued.[59] A new meeting with the government was scheduled
for the end of that month. Herschel was eager to prepare a solid proposal for the meeting. He asked Lloyd and Sabine to draw up a written
account of the probable material and expense for the proposed observations on land and the *project de voyage* for the expedition, in a form
they could hand into the government at the official interview. "I think
a *viva voce* statement of our ideas will hardly suffice" he cautioned.[60]

By the time of this formal meeting with Lord Melbourne, Herschel
was less sanguine. Neither Lord Melbourne nor Thomas Spring-Rice,
the Chancellor of the Exchequer, seemed committal. Minto was also
stubborn, and Herschel feared that the expedition would not be ready
by spring. To Sabine, he admitted "indeed now it would be very difficult to make all the necessary preliminary arrangements with due consideration," yet he still held out hope.[61] But Herschel's optimism did
not last for long.[62]

---

56. Herschel to Minto, November 5, 1838. (RS HS 21.268).
57. Ross to Sabine, November 5, 1838. (PRO BJ3/16/64).
58. Fox to Sabine, November 14, 1838. (PRO BJ3/19/17–18).
59. Lloyd to Herschel, November 13, 1838. (RS HS 11.266).
60. Herschel to Lloyd, November 16, 1838. (RS HS 21.270).
61. Herschel to Sabine, December 2, 1838. (PRO BJ3/26/13).
62. Writing to Gauss, he commented, "What may be the success of these ap-
    plications, it is impossible to say but I have great hopes that something
    will be done worthy of this country." Herschel to Gauss, December 3,
    1838. (RS HS 21.271).

The fatal crack in the lobby came on December 3. Sabine, Ross and Beaufort, collectively representing the naval side of the crusade, met at the admiralty. Realizing that the project could fail if they continued to insist upon the observatories, they were forced to make a decision: abandon the observatories in favor of the expedition or see both parts of the crusade face an increasingly uncertain future. Given such a choice these nautical men decided for the expedition. Sabine was chosen to write to Herschel and inform him of their decision. He apologized for troubling Herschel again after the efforts he had made for the crusade, but was "bound by the desire of my two friends."

> If the expedition be not appointed for the next spring, the labours of yourself, and of your worthy colleagues, in the deputation, will, as respects the accomplishment of the travel of the enquiry, be chiefly lost. I will not reurge that all might be ready by the Spring, both for the expedition & the observations, tho' I continue fully of the opinion, but no one can question that *the branch of research for which the expedition is designed needs no delay.*[63]

Sabine promised that the idea of the observatories could be revisited later, but insisted that the expedition must go out now, alone if necessary. This course of action, the three believed, was all the government was prepared to sanction at the time. Indeed, Ross had heard from a friend at the admiralty on December 2 that the board was certainly in favor of a southern expedition.

The old theoretical differences now threatened to reemerge and divide the lobby by separating the expedition from the stations. For Sabine, Ross and other supporters of Hansteen's theory, the expedition to find the southern poles was the most important part of the crusade and the stations had always been secondary. As men used to sea voyages, they held that all of the necessary observations could be made at sea, and that there was no need for continuous observations at fixed points. Indeed, they believed fixed observatories could be useless, fearing the possibility of "station error" caused by locating a station too close to an iron deposit or other magnetic disturbance. This source of error presented a practical difficulty to the determination of the elements of the theory of terrestrial magnetism from exact observations at a few chosen positions on the globe.[64] Ross later praised the value of observations made aboard ships, as opposed to those made at terrestrial stations. "I feel satisfied that on board ship much of the disturbing influence can be done away with, and the exact value of the remainder determined" he claimed. "Can we say the same of observations on shore, even under the most

63. Sabine to Herschel, December 3, 1838. (RS HS 15.25).
64. "Report of a Committee." *BAAS Report* 11 (1841), 40.

favourable circumstances?"[65] Sabine and Ross also favored a national voyage that emphasized Britain's role in discovering magnetic theory rather than linking British stations to the existing series of continental ones.

For Herschel, Lloyd and supporters of Gauss's theory, the temporary collection of data by sea was less important than continuous collection at fixed points, the exact location of which could be known. This required simultaneous observations at many points, not just data along the line of one expedition. Simultaneous observations in every part of the globe were most desirable.[66] Gauss's theory could only be based upon stations that were spread out. Each station's value depending upon how distant it was from those at which they already possessed observations.[67] Herschel held that the British project should be part of an international venture, in cooperation with other European states, to find a geomagnetic theory.[68] Herschel also felt that the results should be available to every age and nation.[69] This approach was part of Herschel's idea of universal science.

Herschel was certainly taken aback by the sudden defection of his allies. Lloyd supported Herschel's position, having earlier warned Sabine that the project must not be "done *by halves.*"[70] Yet Lloyd was in Dublin and could provide little immediate aid. Fearing that the observatories would be eliminated, Herschel realized that this opportunity could be his only chance to establish stations on such a scale. He needed a powerful ally with government connections. Thus he turned to Spencer Joshua Alwyne Compton, second Marquis of Northampton and president of the Royal Society.[71]

THE ROYAL SOCIETY

If there was a "celebrity spokesman" for the Magnetic Crusade it was the second Marquis of Northampton (1790–1851). Already well known in scientific circles, he had even served briefly as president of the British Association in 1836. The Marquis seemed to be the

65. Ross to Sabine, February 9, 1840. Reprinted in Edward Sabine, "Contributions to Terrestrial Magnetism." *Philosophical Transactions* 130 (1840), 155.
66. Herschel to Jenkins, April 30, 1839. (UTX 1054:202).
67. Sabine to Herschel, July 2, 1839. (RS HS 15.44).
68. Herschel to Lloyd, August 7, 1839. (RS HS 22.25).
69. Herschel to Sabine, January 28, 1841. (RS HS 15.121).
70. Lloyd to Sabine, October 29, 1838. (PRO BJ3/8/52). Emphasis in original.
71. Other scientists, such as Robert Murchison, also appealed to Northampton at this time to press the government for action. Murchison to Sabine, December 16, 1838. (PRO BJ3/26/15).

natural place for Herschel to turn when in need.[72] As president of the Royal Society from 1838 to 1848, he had the influence of the oldest scientific body in the land behind him. He was connected to government and personally knew several prominent ministers, including the Chancellor of the Exchequer, Thomas Spring-Rice.[73] He moved both in scientific and political circles, giving him an important (if sometimes difficult) position.[74] Unlike Humboldt or Herschel he was not a scientific savant. Yet though Northampton may not have known much about geomagnetism (Sabine later declared "that Lord Northampton's knowledge goes not a jot beyond his [year end] address"[75] to the Royal Society), he proved himself to be a very willing ally of the crusade.

The problems encountered with the admiralty concerning the fixed observatories convinced Herschel that he required the full assistance of the Royal Society in achieving success for both sides of the magnetic project. While the initial idea for the crusade may have originated in the British Association, it always had some connections with the Royal Society, both because of the society's past interest in similar projects and because there were so many mutual members of both bodies. It was therefore not unnatural that in times of need the members of the lobby, though officially speaking for the British Association, would turn to its "rival." As will be seen, the Magnetic Crusade was an example of a project in which the two scientific societies worked together closely, rather than competing with one another.

Indeed, the idea of appealing to the Royal Society for aid had already been mooted. In November 1838 Lloyd had suggested that the British Association should coordinate with the Royal Society on the issue of the fixed observatories, since the Royal Society had been investigating

72. Shortly before, Herschel had described Northampton as "a most excellent President for the Royal Society." Herschel to Caroline, November 26, 1838. (UTX 1058:583).
73. Herschel to Sabine, December 16, 1838. (PRO BJ3/26/15). Spring-Rice (1790–1866) was a Whig MP for Cambridge who held several positions in Melbourne's governments. He was Chancellor from 1835–1839, when he was elevated to the House of Lords as Baron Monteagle. DNB.
74. Herschel would later describe the contradictions of Northampton's role as president: "We *cannot* have a President who shall be at once a man of that personal influence with men in power which only rank...can bestow—a man of general science and eminence in some one of the great departments of it—and who will give up *his whole time and attention* to the business of the Society. To preserve the Soc[ie]ty in good working action the last condition is indispensable—but we need never hope to see it realised. No man of any emminence and who has his own pursuits & concerns to attend to can be expected to do it. We must therefore abandon the desireable—and descend to the practicable." Herschel to Sabine, December 15, 1839. (RS HS 15.74).
75. Sabine to Lloyd, December 20, 1839. (RS Te #80).

that line of research since Humboldt's letter of 1836.[76] Minto had also
advised Herschel that the established Royal Society had more weight
than the relatively new British Association in lobbying for the crusade.[77] Just after his first meeting with Melbourne in November,
Herschel was sure that he could get the support of the Royal Society,
should it be needed.[78] The physics and meteorology committee of the
Royal Society, which Herschel chaired, took up the question of the
Magnetic Crusade that same month.[79] By December, this committee
had agreed to forward the matter to the Council of the Royal Society
for a recommendation to the government in support of the British
Association's application.[80]

The lobby was pressing for two distinct goals that formed comple-
mentary parts of the crusade: First, an Antarctic expedition to survey
the southern hemisphere and second, the establishment of fixed observatories in British colonies to take regular readings to supplement the
survey and provide a long-term series of observations. The two halves
of the crusade faced different paths before they finally received approval
and it required considerable effort to prevent them from being split
apart, an action which might potentially sacrifice one in favor of the
other. By December 15, Herschel had recorded that Ross was having
some success on behalf of the Antarctic expedition.[81] Yet while Ross
could be counted on to back the naval part of the expedition, Herschel
still needed support for the fixed observatories. This aid he received in
part from Dublin when Lloyd finally weighed in on Herschel's side
(against Sabine) in favor of keeping the fixed observatories as an integral
part of the crusade.[82] Another meeting with Melbourne was scheduled
before Christmas to argue for an early departure of the expedition, with
the observatories intact. But Herschel still required an ally to help secure the creation of his global system of observatories.

Now in late December, Herschel turned directly to appeal to the
President of the Royal Society, Lord Northampton, for the help of that

---

76. Lloyd to Herschel, November 17, 1838. (RS HS 11.268).
77. Diary: November 23, 1838.
78. Diary: November 29, 1838.
79. Diary: November 30, 1838.
80. Diary: December 15, 1838. Friendly pushes Beaufort's role in the lobby at
    this stage, pointing out that Beaufort was "strategically placed on [the]
    Physics and Meteorology Committee," without even mentioning that
    Herschel was its chair. Friendly, 290.
81. Diary: December 15, 1838.
82. "I believe that it will be *possible* to provide the material for the fixed ob-
    servations by next spring...I cannot bring myself to accede to your notion
    of *separating* the two parts of this grand project. However, if you succeed
    with government, I need not say that I shall do my best to fulfil my parts
    connected with the undertaking." Lloyd to Sabine, December 24, 1838.
    (PRO BJ3/8/59).

august body for the final phase of the lobby.[83] "There seems every reason to believe that if the representations of the Council of the R[oyal] S[ociety] on the subject of Magnetic operations be not immediately and favorably replied to the whole affair will be a *coup manqué*" Herschel warned, "not merely for the present, but absolutely, since there cannot but be the greatest difficulty in bringing the matter on any future occasion up to the point it has now obtained."[84] Fortunately Northampton was favorable to the project, and promised to express his opinion of the great importance of the Crusade both in a scientific and national point of view to the government.[85] With newfound support, the lobby was ready for the final assault on the admiralty in support of the crusade. In late December, Captain Beaufort recommended that Herschel "fire another shot at Minto."[86] In early 1839 Northampton was to meet with the Cabinet to argue the case for the lobby.[87]

The addition of the Royal Society's support for the British Association proposal also had effects within the lobby. By the beginning of 1839 Sabine had come back over to support Herschel's side in the debate over the priority of the expedition and the stations. Sabine thought that the authority that the Royal Society brought to the lobby gave it a new, professional feel.[88] He seemed to be more optimistic about the future of the crusade and appeared impressed by the Royal Society's intervention on behalf of the observatories. To Lloyd, Sabine commented that he was satisfied that everything had been done that could be expected and that he was willing to accept the decision of the government—whatever it would be.[89] This new backing from the Royal Society encouraged Sabine to expect a positive reply from the government. In the first week of 1839 he wrote to Ross that a new deputation to Lord Melbourne was to occur on January 5, when they should learn the fate of the expedition for the present year. Sabine's opinion was that they would agree to it at once.[90] Whewell was also cautiously optimistic at this time.[91]

---

83. Diary: December 22, 1838.
84. Herschel to Northampton, December 23, 1838. (RS HS 21.273).
85. Northampton to Herschel, December 31, 1838. (UTX 1087:400).
86. Beaufort to Herschel, December 24, 1838. (UTX 1087:54).
87. Northampton to Herschel, December 31, 1838. (UTX 1087:400).
88. Sabine to Ross, January 3, 1839. (PRO BJ2/13/7).
89. Sabine to Lloyd, January 3, 1839. (RS Te #59).
90. Sabine to Ross, January 3, 1839. (PRO BJ2/13/7).
91. To Quetelet he wrote "we are endeavouring to induce our Government to send out an expedition towards the South Pole to determine the present position of the Southern Magnetic Pole, and also to establish several permanent magnetic observatories in order to observe the simultaneous changes discovered by Gauss and his friends." Whewell to Quetelet, January 17, 1839. (WC).

The support of the Royal Society seemed to bode well for the future of the crusade. Comments were also positive from outside of the lobby. Robert Fox wrote to Sabine that the appeal of the Royal Society to the government in the cause of terrestrial magnetism was too strong to be easily refused.[92] In addition, Sabine found that the lobby had more external help. He informed Herschel that commercial interests were now joining the call for the crusade. South Sea merchants, trading in oil, were getting up a memorial arguing the advantage to their commercial interests of good geographical and magnetic charts of the southern ocean.[93] This commercial approach gave the lobby a new aspect and a new source of support going into the home stretch.[94]

However, all did not go as well as the lobby hoped. When Northampton was unable to attend the Cabinet meeting with Melbourne, government delayed action even further. Sabine relayed a report of the meeting (via Beaufort) to Herschel:

> Captain Beaufort informs me one of the Cabinet has said "the recommendations of the R. S. have been brought before the Cabinet, but we have decided that they should stand over. Lord Northampton the President did not come up with the deputation." Of course that is only a excuse but I am sorry that they have found one in the supposed lukewarmness of the President.[95]

The support of the Royal Society turned out to be insufficient without the presence of its President. The constant bureaucratic delays now began to threaten the timetable for the naval expedition, which the lobby had hoped would leave by spring. By February, Ross was venting his frustration:

> I have delayed writing my own [statement] day after day in the hope that this wavering trifling government of ours would feel themselves bound to make up their minds regarding the Antarctic Voyage, that I might have it in my power to acquaint you with their determination, but the proper season for active operation has been surely frittered away by audience after audience, deputations & petitions & still *nothing is decided!*[96]

It was in February 1839 that the lobby finally had its next meeting with Melbourne and this time Northampton (prepared by Sabine in advance) was in attendance. Sabine went into London to meet with

92. Fox to Sabine, January 22, 1839. (BJ3/19/34).
93. There was also "a memorial from the Geographical Society...in preparation." Sabine to Herschel, n. d. [January 1839]. (RS HS 15.322).
94. The lobby for the Wilkes expedition in the United States was largely propelled by commercial interests which at times were a disadvantage. See Chapter IV.
95. Sabine to Herschel, n. d. [January 1839]. (RS HS 15.322).
96. Ross to [...], February 6, 1839. (PRO BJ2/5/2). According to the finding aid, this letter is to Ross's wife.

Herschel, in case he should wish to "go *thro' the form* of meeting the Secretaries who are appointed to him to draw up the report." Sabine's comment that "the report I doubt not will be, as usual, ready written in his [Herschel's] waistcoat pocket" again demonstrates the influence which Herschel and his ideas had in the lobby at this point. By now there was no doubt that the lobby was being run by Herschel and his allies.

Sabine had emphasized the points which he wished Northampton to make to Lord Melbourne at the time of their meeting, that the expedition and the fixed observatories should be approved together so that they could set out at the same time.[97] He held that if acted upon quickly, the expedition could be ready to sail in three to four months. Sabine confided to Lloyd that he believed such a push would not endanger the observatories, as he held that once the expedition was approved, it was easy to add them into the project.[98]

The central problem continued to be Lord Minto and his qualms about the timing of the establishment of the fixed observatories.[99] Even with Melbourne's approval, the lobby still had to deal with the admiralty, whose job it was to actually carry out the instructions.[100] In the end, it was Ross who managed to find a way around the objections. Ross knew that Lord Minto was the chief cause of the delay, because he wanted the fixed observatories to be established first, then have the expedition sent out.[101] Meeting Minto at the admiralty, Ross pointed out to him the advantages in savings of time and money if the different observatories were established by the crusade on the way out.[102] Ross

---

97. Sabine to Northampton, February 8, 1839. (PRO BJ2/13/9).
98. "On the occasion of his [Northampton's] presenting the recommendation of the R S [he ought] to press *that the expedition may be fixed for the next Spring*, without reference to the question of the preparation or otherwise of the instruments and observers for the fixed observatories. If the one is planted the other will be *sore* to follow. But we must make a strong push to get both out at the same time." Sabine to Lloyd, n. d. [February 1839]. (RS Te #62).
99. Appropriately for a man who caused so much trouble for the inductive aims of the crusade, Minto was a great admirer of David Hume, a fellow Scotsman. Commenting on a recent biography of Hume, he wrote "I trust that the book will be read and appreciated as it deserves and that our neighbors on the south side of the Border especially will learn from it to do justice to the memory of one of the most profound and original thinkers, as well as one of the most amiable and virtuous men, that our country has produced." Minto to [...], April 10, 1846. (NLS, MS 3005 fol.s 126–127).
100. Whewell to Sabine, n. d. [February 1839]. (PRO BJ2/13/8).
101. Ross to [...], n. d. [February 1839]. Although undated, this letter refers to the same conversation with Minto that Ross mentioned in a letter dated February 11, 1839 (PRO BJ2/5/31).
102. Ross to Sabine, February 11, 1839. (PRO BJ3/16/78).

believed his argument to have been successful. Some other questions of minor importance were advanced by Minto and successfully countered by Ross so that before they parted Ross was convinced that Minto "most strongly concurred with that said and that he was most strongly in favor of the enterprise."[103]

Minto's remaining objections concerning the observatories were finally overridden by simply removing the responsibility from his department altogether. Ross reported that Minto "threw the fixed observations overboard" saying that he would have nothing to do with them.[104] Thus after months of debate, the issue seemed to be put to rest. It was only at this point in March 1839 that Sabine suggested using officers of the artillery to man the fixed stations, thus removing responsibility from the admiralty (and Lord Minto) altogether. "We could not do better than look to the Corps," Sabine assured Herschel, "no other difficulty could remain but that of finding the individuals themselves."[105] One week later, Sir Alexander Dickson, the adjunct general of artillery, and Colonel Fox, the master general's secretary, reported that the authorities at the Ordnance supported the proposal of employing officers of the artillery or engineers at the fixed observatories.[106] With the Artillery's support replacing that of the admiralty and with a new pool of potential observers, the pieces were all in place to launch the crusade.

By the spring of 1839, the lobby seemed to have been on the verge of success. In March, Minto announced that the government had officially approved the nautical part of the crusade.[107] Herschel was optimistic that

> we may now count on both branches of the Magnetic undertaking as resolved on. The naval certainly, and the orders is understood to be issued for selecting the vessel or vessels—Ross to command—The fixed observatories also have the assent of L[o]rd Melbourne & the Ch[ancello]r of the Excheq[uer] and are only not definitively declared on the score of communication understood to be merely formal. In all probability both branches will be officially announced at the next council of the R[oyal] S[ociety].[108]

There was some negotiating to do about the details of the expedition, which was still under the admiralty. For example, Ross feared that Minto was unaware of his desire for two ships.[109] These objections

---

103. Ross to [...], n. d. [February 1839]. (PRO BJ2/5/31).
104. Ross to Sabine, March 14, 1839. (PRO BJ3/16/82).
105. Sabine to Herschel, March 15, 1839. (RS HS 15.28).
106. Sabine to Herschel, March 21, 1839. (RS HS 15.31).
107. Diary: March 11, 1839; Herschel to Whewell, March [11], 1839. (RS HS 22.41).
108. Herschel to [Sabine], March 13, 1839. (PRO BJ3/26/21).
109. Ross to Sabine, March 14, 1839. (PRO BJ3/16/82).

were finally worked out at a dinner party given by Lord Northampton on the night of March 16. With this successful behind-the-scenes lobbying, Lord Minto, Northampton, Herschel and Ross made an appointment to meet, choose the ships, and settle all necessary preliminaries.[110] Shortly thereafter Sabine reported success for the expedition, although he cautioned that the observatories (now no longer under the admiralty) still required final approval.[111]

Despite recent progress the future of the fixed stations was still undecided, a position that left some members of the lobby uneasy. Herschel knew that, while the nautical part of the Magnetic Expedition was officially recognized, the Council of the Royal Society had yet to receive any written communication or any distinct authority to take any steps on the fixed observatories.[112] Although there had been an agreement to establish the observatories, a great number of preparations still needed to be carried out before the expedition was launched. Instruments had to be ordered and officers chosen and trained by the scientific community to man the stations. Arrangements also had to be made to coordinate the observations at the stations with those carried out by the expedition and others being conducted in Europe at the same time. Until the orders and funding were obtained to carry out these preparations, the fate of the observatories still hung in the balance. Herschel now feared that the naval expedition would be detained by the delay in approval for the observatories.[113] He was anxious that

> at present we stand rather awkwardly. Nothing is yet officially referred to the R[oyal] S[ociety] for consideration or action but on that our understanding will be speedily come to. However until that is come to we can *do* nothing tho' we may consider & prepare much.[114]

Now at the end of March, a final obstacle loomed.

By March 28, the admiralty had assigned the ships *Erebus* and *Terror* for the Antarctic expedition.[115] But a note from Ross on March

---

110. Sabine to Lloyd, [March 17, 1839]. (RS Te #66).
111. "The expedition is ordered. The Terror & Erebus are selected as the ships. Ross to command. They are directed to be ready for sea in *June*, but we may view early in *August* as the real time of departure. The fixed observatories not being in Lord Minto's department, another reference has been necessary to Lord Melbourne, which was made on Wednesday, where it was fully understood from him & the Chancellor of the Exchequer, that the order for them also is only delayed by a point of form, and that the official certification may be expected at the Council of the Royal Society which meets next Thursday, [March] the 21st." Sabine to Lloyd, n. d. [March 1839]. (RS Te #61).
112. Herschel to Whewell, n. d. [March 1839]. (RS HS 22.41).
113. Herschel to Ross, March 29, 1839. (RS Sa MS258).
114. Herschel to Sabine, March 19, 1839. (PRO BJ3/26/22–3).
115. Ross to Herschel, March 28, 1839. (UTX 1093:461).

30 indicated that the naval side of the project might once again be wavering in its support of the observatories. Captain Beaufort had warned Herschel that it was not advisable to make any further application to the Admiralty for instruments for the observatories at present. Already, preparations for the observatories were holding up the expedition. Ross felt that "it will be a sad day" if a delay in the establishment of the fixed observatories were seen as grounds for delay in the naval portion also.[116] The threat that the observatories might be sacrificed for the expedition again moved Herschel into action and he undertook a marathon of writing (at least nine letters) over the night of March 31, 1839. To Whewell he wrote that there had been a verbal understanding that the observatories would be approved, "but nothing *is* ordered." Herschel feared that the stations, which he had always seen as an equal part of the crusade that should be founded on the outgoing voyage, would not receive as much support now that the expedition had been approved and might flounder for lack of equipment—"Here therefore is a *causas foederis*."[117] To Sabine and Ross he wrote that as the land part of the magnetic undertaking was still in a very anomalous state, he greatly feared that the naval part might ultimately be delayed, "a most disastrous commencement" which he sought to avert.[118] While Herschel warned of a delay in the expedition, clearly he was more concerned with the fate of the stations.

Herschel was ready to use whatever means necessary to achieve his goal. In an appeal to Whewell he maintained that:

> we want no more *formal deputations or interviews*, but somebody wants a jog. Is not Peacock a personal friend of Spring-Rice and could he not suggest to him by letter or word the extreme awkwardness and inconvenience of the present state of things?[119]

Herschel pulled out his own personal connections for the task as well. Once again he turned to Northampton to bring the full weight of the Royal Society to bear on the government to give official approval for the stations in addition to the expedition. Herschel pointed out that the official orders for the equipment for the Antarctic expedition did not extend to the fixed observatories and he feared the possibility that the expedition would be equipped and the ships ready to sail before those arrangements could be made, unless the organization of the fixed observatories was placed on an equal footing. He suggested that a note from Northampton to Lord Melbourne was all that was required

116. Ross to Herschel, March 30, 1839. (UTX 1093:462).
117. Herschel to Whewell, March 31, 1839. (RS HS 22.6).
118. Herschel to Sabine, [March 31,1839]. (PRO BJ3/26/28–9); Herschel to Ross, April 5, 1839. (RS Sa MS258).
119. Herschel to Whewell, March 31, 1839. (RS HS 22.6).

to accomplish this task.[120] Another friendly note came from Peacock to Spring-Rice, the Chancellor of the Exchequer, ("a personal friend of his") on the same subject.[121]

In early April, Herschel learned that Melbourne had given the duty of establishing the stations to Spring-Rice but that he had not seen to it yet.[122] Nevertheless Herschel optimistically told Lloyd to go ahead and order the instruments for Tasmania, and to be prepared to do the same with the other observatories once official confirmation came.[123] Herschel's advice was not premature. Sabine saw Spring-Rice on the morning of April 10 and found that the orders for the observatories had gone out that day.[124] The next day Sir Hussey Vivian finally received orders at the Ordnance for stations at Canada, the Cape and St. Helena.[125] Herschel wrote to Sabine to express his pleasure that the stations had finally been approved.[126] On April 12, Sabine informed Lloyd that the Royal Society had received the government's orders for the establishment of the colonial observatories.[127] Herschel lost no time in requesting that Lloyd order all of the remaining instruments for the observatories and draw up a series of instructions along with skeleton forms of observation and reduction for all the instruments to be used in each observatory to simplify matters for the observers.[128] The Magnetic Crusade was under way.[129]

---

120. Herschel to Northampton, [April 1, 1839]. (RS HS 22.74). This important letter has not received much attention, perhaps because it was misdated simply "1840" in the Herschel correspondence, apparently placing it well after the stations were approved. Internal evidence from the letter (such as Herschel's upcoming trip to Kent and his report of a visit to the Admiralty that morning) when compared with Herschel's diary indicates that it must have been written on the morning of April 1, 1839.
121. Herschel to Sabine, April 5, 1839. (PRO BJ3/26/24–5).
122. Sabine to Herschel, April 4, 1839. (RS HS 15.33).
123. Herschel to Sabine, April 9, 1839. (PRO BJ3/26/30–1).
124. Sabine to Herschel, April 10, 1839. (RS HS 15.35).
125. Sabine to Herschel, April 11, 1839. (RS HS 15.36).
126. Herschel to Sabine, April 12, 1839. (RS HS 22.9).
127. Sabine to Lloyd, April 12, 1839. (RS Te #67).
128. Lloyd to Sabine, April 16, 1839. (PRO BJ3/8/89); Herschel to Lloyd, April 21, 1839. (RS HS 22.10). The funding for the instruments was at issue during this period. The British Association had agreed to contribute £400 for the purchase of scientific instruments for the crusade, but this would not have been nearly enough for the expedition and all of the observatories. Fortunately in the end, the government was convinced to pick up the entire cost. Lloyd to Sabine, April 27, 1839. (PRO BJ3/9/6).
129. Susan Faye Cannon suggests that the temporary fall of Melbourne's government in the spring of 1839 may have threatened the future of the crusade, as Robert Peel's incoming ministry was not believed to be as sympathetic to science. Susan Faye Cannon, *Science in Culture* (New York: Dawson, 1978), 251; Friendly, 291.

Sabine outlined the plan of the expedition for Herschel in July. After sailing to St. Helena and the Cape, Ross would proceed toward Australia, establish an observatory at Tasmania, and then sail south toward the Antarctic.[130] Although Ross would be exploring unknown regions, Sabine was adamant that the crusade was not an "expedition or a *voyage of discovery* to the Antarctic." He took issue with one such description of the crusade, arguing that it "both departs from fact, and does so little justice to what the Royal Society really did recommend." He suggested "A Naval Expedition for Researches in Physical Science in the Southern Hemisphere" as a more suitable title.[131]

As the time for departure grew near, members of the lobby expressed their good wishes to Ross. Beaufort wrote:

> expressing to you the high hopes...in common with the whole country. I indulge respecting the successful, I might say glorious results of your voyage and, I am almost ashamed to confess, the still more ardent hopes in which I also allow myself to indulge for your personal share of that glory and that success.[132]

Herschel, realizing that he could not see Ross before his departure, also wrote to wish him every possible success on the voyage.[133] Ross's departure in September marked the culmination of over a year's work within the British Association and the Royal Society.

EAST INDIA COMPANY

Running parallel to the activities of the lobby for the Magnetic Crusade and in some ways connected to it was a separate effort within the East India Company. By the nineteenth century the company, once a great trading enterprise, had become the face of British imperialism in Asia. Trading posts had become colonies and the company acted as the representative of the British Crown until the time of the Sepoy Mutiny (1857), when the state took over full control. The company had also become involved in scientific interests as well, especially surveys which mapped their vast domains and located potential resources. Early in 1838, Captain Thomas Jervis had tried to convince the court of directors to fund a geographic survey of central Asia. Jervis was an associate of Sabine, who he hoped to induce to come to India with him.

Although the East India Company turned down Jervis's proposal in July 1838, he decided to try again in August, just as the lobby for the

---

130. Sabine to Herschel, July 2, 1839. (RS HS 15.44).
131. Sabine to Herschel, August 9, 1839. (RS HS 15.55).
132. Beaufort to Ross, September 26, 1839. (PRO BJ2/3/11).
133. Herschel to Ross, n. d. [1839]. (PRO BJ2/6/65).

Magnetic Crusade was getting started. Indeed, Jervis's effort probably benefited from the attention raised by the British Association for their magnetic project. Writing to James Melville, Jervis noted that some scientific topics under investigation in Europe could be carried out in India at a comparably small expense. If the instruments were provided, the observations could be conducted simultaneously on the same principles as those in Europe. He went on to suggest a magnetic survey in India, requesting funds for "apparatus for three fixed magnetic observatories at Madras, Bombay and the Surveyor General's office; [and] Apparatus for six traveling observatories with meteorological instruments."[134] The addition of these observatories made this proposal different from Jervis's previous one.

Members of the magnetic lobby became involved in the East Indian venture as well. By August 1838 the lords commissioners of the admiralty had referred Jervis's new proposal to a committee consisting of Herschel, Francis Baily, Whewell and Airy, which it was hoped would serve as a sort of scientific advisory board for the East India Company on this and other matters.[135] The participation of some of the same prominent scientific figures in both projects ensured that there would be a degree of cooperation. Herschel and Whewell provided the main connection between the two projects, which coordinated efforts throughout British Asia. In March 1839, Lloyd separately approached the surveyor general of Ceylon with the possibility of establishing an observatory there.[136] By that same month, Jervis had ordered four sets of instruments for the Indian observatories, indicating that the East India Company had given tentative approval to the project.[137] This action had rendered the Ceylon observatory "less necessary" and Sabine and Lloyd now began discussing the best emplacements for the Indian observatories. "Ought we not to have one of the 4 high up on the Himalayas?" queried Sabine. "One Madras, one Bombay, one Simla and one where else?"[138]

The full scope of the East India Company's commitment was to become apparent that spring. The company was prepared to extend the initial magnetic observations to include other geophysical phenomena as well. At the end of March 1839, Jervis could assure Lloyd that orders had gone out for tidal, meteorological and magnetic observations at twenty-five stations under company control throughout Asia.[139] The full list of the equipment for the East India Company stations gives an

134. Jervis to Melville. August 15, 1838. (BL Add. 34649 fol. 119).
135. Herschel, though was wary of this new demand upon his time. Herschel to Robertson, August 6, 1838. (RS HS 21.257).
136. Lloyd to Sabine, March 10, 1839. (PRO BJ3/8/70).
137. Sabine to Herschel, March 15, 1839. (RS HS 15.28).
138. Sabine to Lloyd, [March 16, 1839]. (RS Te #61).
139. Jervis to Lloyd, March 28, 1839. (RS Te #63).

idea of the character of the observations (and the expense that the company was willing to incur): Sixteen Osler anemometers (£640); thirty-two barometers; twenty-three chronometers (£1,035); ninety-nine varieties of transits (£360) and magnometers (£665), or £2,700 in all.[140] In September 1839, the secretary to the government of India, Henry Prinsep, sent out an order to conduct simultaneous storm observations in as many stations in India as possible.[141] Given the ambitious extent of the project, Lloyd could not "help thinking that the Company's scientific advisors [we]re *overdoing* the business."[142] Sabine, however, felt that the East India Company observatories would afford interesting comparisons across Asia.[143]

Herschel himself became directly involved in the East India Company's efforts to set up observatories in India when Jervis appealed to him for help with the officers of the company. Lord Northampton had omitted in writing to Sir Richard Jenkins (chairman) to point out the urgent necessity of training officers in the manipulation of the instruments and the course of observations to be made.[144] As a result, Herschel himself wrote to Jenkins, stressing the importance of the East India Company's contribution and again appealing to nationalism. He hoped that the East India Company was sufficiently impressed with the importance to both practical and theoretical science of the subject, and as the present opportunity was unlikely to occur again they might participate in an undertaking Herschel saw as having "an eminently national character."[145] As a result of Herschel's request, the departure of the observing officers for India was delayed by a month so that they could undergo further training with the instruments they were to use.[146] During that spring, Herschel had also been consulted by John Hobhouse from the India Board regarding the plan for the observatories in India, and to whom he recommended an observatory on Socrata.[147]

By November 1839 the officers who were to man the Indian observatories had been sent to Dublin for training with Lloyd. Although there

---

140. Lloyd was convinced that "this is some of Jervis's doings." Lloyd to Sabine, December 11, 1839. (PRO BJ3/9/72–5). Compare the British Association grant of £400 for instruments and Lloyd's estimate of £1,500–£2,000 which "would difray [sic] all the expenses of each observatory for the *three* years of its work, including the extra pay usually given to officers employed in such service." Lloyd to Herschel, November 13, 1838. (RS HS 11.266).
141. H. T. Prinsep, "Notification," September 11, 1839. Reprinted in Matthew Maury, *Explanations and Sailing Directions to accompany the wind and current charts*. (Washington: Harris, 1858), 26.
142. Lloyd to Sabine, May 30, 1839. (PRO BJ3/9/24).
143. Sabine to Lloyd, July 5, 1839. (RS Te #70).
144. Jervis to Herschel, April 26, 1839. (UTX 1087:298.2).
145. Herschel to Jenkins, April 30, 1839. (UTX 1054:202).
146. Melville to Herschel, June 15, 1839. (UTX 1083:951).
147. Herschel to Hobhouse, June 4, 1839. (UTX 1054:194).

was still some dispute as to the locations of the stations (Lloyd held that "Madras [wa]s an *unhealthy* station"), the East Indian Company directors were anxious to assist, and wished for more information on the project.[148] Although the company had appointed the officers in charge of the observations rather than the Royal Society, the company assured the scientists that they would only choose duly qualified officers who were "zealous in the cause."[149]

Eventually three magnetic stations were set up in India: at Bombay, Madras and in the Himalayas. The observers were sent out at the end of November. Their observations were sent back to Lloyd in Dublin just like those from the other colonial observatories. Lloyd was convinced that the company had taken up the matter in a businesslike manner, and was particularly impressed with the importance of having these observatories in cooperation with those already established.[150] The East India Company's efforts greatly expanded those of the Magnetic Crusade, allowing the project to cover an even greater part of the world than Herschel and the lobby had initially imagined. With the addition of the East India Company, the major elements of the British state (military, scientific and imperial) were all cooperating in the magnetic venture.

▒ "THIS IS HERSCHEL'S DOING"

Herschel's philosophical system and his influence in the lobby helped to push the stations into an equal position with the expedition. Twice he went out of his way to maintain their position even when the lobby was having troubles. Herschel's decision to involve the Royal Society in the lobby proved a decisive element. Later, Herschel maintained that "without Lord N[orthampton]'s personal intervention I feel, indeed I may almost say I know, that the magnetic project would not have been adopted by the Ministry."[151] The intervention of the Royal Society also helped to give the lobby a more unified appearance, now as the work of the whole scientific establishment rather than just one group. The addition of the East India Company stations completed the national appearance of the project. With the participation of the East India Company the Magnetic Crusade had "for the first time moved to a definite proposal of a national character."[152]

---

148. Lloyd to Sabine, November 18, 1839. (PRO BJ3/9/64–5).
149. Lubbock to Herschel, August 9, 1839. (UTX 1087:340).
150. Lloyd to Sabine, November 30, 1839. (PRO BJ3/9/68).
151. Herschel to Sabine, December 15, 1839. (PRO BJ3/26/102–4).
152. Draft of a reply written by Herschel on Jervis to Herschel, April 28, 1839. (UTX 1087:298.2).

Indeed, the Magnetic Crusade, while it originated with the British Association, became just as much a project of the Royal Society. The recommendation for the East India observatories was made on behalf of the Royal Society, not as part of the efforts of the British Association.[153] In December 1838 Northampton had stated his view that the lobby for the crusade expressed "the opinion of the British Association as well as the R[oyal] S[ociety]."[154] When the official orders went out from the government for the establishment of the observatories, it was to the Council of the Royal Society, not the British Association.[155] The two bodies, which shared many members, became so enmeshed in the activities of the lobby that it was sometimes difficult to tell which had recommended what. In April 1839, for example, Sabine pointed out to Herschel that a committee of the Royal Society had actually authorized the expenditure of funds that had been set aside by the British Association![156] Herschel in turn pointed out that since all of the members of the British Association committee which had allocated the funds in the first place were also fellows of the Royal Society "either on the council or on one or other of the comm[itt]ees and being all fully impressed with a sense of identity of object there is no possibility of any misunderstanding on this point."[157] Here there was clearly no division between the efforts of the two bodies.

The establishment of the stations not only helped to begin the global data collection for which Herschel had been pushing since the previous year, it also helped to tie Britain into the international community of magnetic observatories. The crusade, however, was more than just an extension of the continental system of observatories. The British plan called for more extensive observations than ever before, made on a scale that dwarfed the existing German *Verien*. Sabine once explained to Lloyd that Herschel had a plan for distributing the work among the various observatories "so that certain of them shall be always on duty, as it were." Nor was the crusade a one-shot voyage that brought back only a limited series of observations that would quickly become out of date. Long after the expedition was over, data continued to be generated by the stations in the British colonies, providing invaluable material from the periphery for theoretical work back at the center of calculation. In later months, Herschel worked to expand these stations beyond the few originally approved by the government. In July 1839, Sabine reported on one of Herschel's latest attempts for an observatory in Egypt, emphasizing

---

153. Sabine to Herschel, February 12, 1845. (UTX 1093:521.2).
154. Northampton to Herschel, December 31, 1838. (UTX 1087:400).
155. Sabine to Herschel, April 11, 1839. (RS HS 15.36).
156. Sabine to Herschel, April 10, 1839. (RS HS 15.35).
157. Herschel to Sabine, May 12, 1839. (PRO BJ3/26/38).

the importance of the role Herschel played. "You are aware that we have written to the pasha of Egypt," he exclaimed to Lloyd. "This is Herschel's doing."[158]

The expansion of the crusade to include fixed observatories also tied the scientific venture more closely to the imperial apparatus of the British state. This element more than any other set the crusade apart from similar ventures, by creating a new relationship between science and state. This interaction benefited the project both by helping to secure government support and extending the number of stations and the places where they could be located. Government backing ensured funding and backing for the scientific project, while the state gained prestige and scientific information on tides, navigation and climate that could benefit imperial expansion. Indeed, the crusade fit nicely into existing imperial policy on science. Already the admiralty was sponsoring geographical surveys of India, and in the fall of 1838 the governors of British colonies were asked to begin keeping records on storms and winds.[159] The fortunate involvement of the East India Company radically changed the scope of the venture. East India stations soon far outnumbered the ones initially planned by the crusaders. The inclusion of the stations changed the nature of the crusade, integrating the system of observations into those conducted in Europe as well as expanding it into the empire.

That the idea for such an extensive system of stations originated with Herschel seems clear. Neither Lloyd nor Sabine was willing to suggest such a plan in August 1838 when the lobby began. The addition of the observatories to the crusade was an act always supported and defended by Herschel. Sabine later came to see them as the more important half despite his affection for a naval expedition and his part in the December 1838 mutiny that had almost succeeded in removing the fixed observatories from the crusade altogether. A year later he admitted to Lloyd that

> I believe with you that the results of the system of fixed observatories will eventually be even more important than those of the naval expedition, and that, particularly, establishment of the observatories will constitute in the view of our successors as of much more consequence than the expedition & its results.[160]

---

158. Sabine to Lloyd, July 5, 1839. (RS Te #70). See Herschel to Northampton, August 1, 1839. (RS HS 22.22): "Resolved—that the President be requested to apply to the Secretary of State for Foreign Affairs for a letter introducing & recommending the magnetic circular of this society to the notice of his highness the pasha of Egypt, to be delivered with that circular to his highness."

159. "Circular to Governors of British Colonies," November 29, 1838. Reprinted in Maury, 27.

160. Sabine to Lloyd, December 20, 1839. (RS Te #80).

Sabine now embraced the observatories despite their lower profile.[161]

Herschel was still operating in the same vein he had been when he had helped to set up meteorological observatories while in Africa, still looking for ways to universalize induction to find physical theories. By setting up observatories that could conduct global observations across significant periods of time, Herschel thought that he could find a solution to deriving general, universal laws from particular data. Herschel's participation in the Magnetic Crusade can be seen as a continuation of the work he had been doing for many years before. Inspired by his belief that universal knowledge was possible and that the weaknesses of induction could be eliminated with enough physical data spread out over time and place, he pursued the goal of founding stations around the world for the purposes of simultaneous physical observations. Herschel's importance in establishing observatories as part of the crusade can clearly be seen from his actions. As late as November 1838, Herschel had insisted that his role in the lobby "must be limited to general advocacy."[162] Yet as the future of the lobby, and especially the stations, became more uncertain in the winter and spring of 1839, Herschel became increasingly active and twice (in December and April) made an especially strong stand in favor of the stations.

In a letter to Arago of October 1839, Herschel spelled out the reasons and motives for his participation in the crusade. Herschel saw this occasion as a truly unique moment in the history of science to extend physical observations on a global scale, an

> opportunity such as may probably never again occur of fixing for future ages—so to speak at a blow...the magnetic data...upon a scale which may be said without exaggeration to embrace the whole globe and which shall spread over a period sufficiently long to give complete room for the elimination of all that is accidental and temporary.

Hoping to gain French assistance in setting up an observatory in Algiers, Herschel described the recent successes. In addition to the stations set up by the expedition, new stations had been founded in India. Applications had been made for others at Aden and Singapore and Herschel hoped that the United States would soon become involved. Britain was now connected to an international system of observing. Herschel's dream and its need for simultaneous, continuous

---

161. "I am fully persuaded of myself, that except by three or four individuals the importance of the magnetic observatories is not yet known or felt in this country; and that *that* is the reason why the branch of the magnetic researches has been so inadequately noticed." Sabine to Herschel, December 16, 1839. (RS HS 15.76).

162. Herschel to Lloyd, November 16, 1838. (RS HS 11.267).

observations around the world was becoming a reality. A station at Algiers would help in the geographic coverage:

> taken in conjunction with London, Paris, St Helena and the Cape it forms an admirable link in a [latitudinal] chain of stations, while if viewed in connexion with Aden, Bombay, Madras and Singapore it would offer a no less advantageous disposition in longitude.[163]

Herschel was still thinking on a global scale. While the crusade may have been sold to the British government as a boon to navigation and commerce, in many ways it represented science's use of the state apparatus for its own ends. For Herschel, the Magnetic Crusade provided the opportunity to fulfill a scientific quest that he had begun years earlier in his *Preliminary Discourse* with its proposal to use stations to collect worldwide data which could be transmitted back to develop new theories. This plan had continued in his 1835 letter to Beaufort concerning physical observatories, and in his 1838 appeal to the admiralty to set up observatories in British colonies.[164] His plans finally came to fruition in the fixed stations established by the Magnetic Crusade, which set up the first of dozens of stations around the world connected back into the continental observatories in a single system. It is thus possible to see at least one half of the Magnetic Crusade as Herschel's own creation.

---

163. Herschel to Arago, October 30, 1839. (UTX 1054:37). Following the British example, a number of other nations established stations for geophysical observations in this period. By 1847, observatories were operating at Algiers, Brussels, Prague, Cadiz, Philadelphia, Cambridge, Cairo, Trevandrum and Lucknow. Whewell, 1857, III:51.
164. Herschel was obviously impressed by his 1835 letter to Beaufort. As late as 1842 he referred Wheatstone to it as a reference for the necessary qualities of physical observatories. Herschel to Wheatstone, June 17, 1842. (RS HS 22.125).

Chapter Four

We Should Also Contribute
Our Mite

In the last chapters we have seen how British scientists were able to negotiate with their existing sociopolitical system to create the crusade. Behind the issue of science and state lies the political environment in which science operates. In different states, diverse political climates existed in which science could either flourish or stagnate. To accentuate the role of politics in the process of creating a scientific project, this chapter evaluates the divergent political processes in 1830s Britain and America that led to the launching of similar Antarctic expeditions. The results point out how the different political climates that scientific proposals faced in those two countries shaped the outcome. Proposals that started out as similar plans could transmute into largely different projects whose success was influenced as much by the politicians who approved the funding as the scientists who crafted the theories. In Britain, the more aristocratic form of government allowed for the development and implementation of projects by scientists working within government without the need to appeal to Parliament or public opinion. The American system, which included a larger degree of popular participation, would be less interested in pure science and more concerned with practical results. Any successful proposal in the republic would also have to pass muster with Congress. These contrasting systems had a noticeable effect on both the type and extent of scientific ventures that could be carried out. In the final analysis, the successful outcome of such projects would depend as much upon the sociopolitical structures that supported them as the scientific foundations upon which they were built.

AMERICAN SCIENCE

The position of the geophysical sciences in the United States was similar to that of Europe in the early nineteenth century. As in Europe, terrestrial magnetism, meteorology and hydrography gained new attention and the Humboldtian approach of series of observations was popular. But unlike in Europe, there were fewer established academic seats or institutions that could carry out and coordinate such observations. Such a situation encouraged the sort of division and democratization of scientific labor suggested by Herschel. "Since the scientific and technical community was so small," notes Nathan Reingold, "such a reduction to practice also presupposed and required an openness of entry into the community. A few gentlemen-scientists could not perform the resulting range of theoretical and applied work."[1] In this period, both geomagnetic and meteorological observations were being made in parts of the country.[2] In New York, Joseph Henry (1797–1878) provides an example of the sort of survey work being done in this period.

Born and raised in Albany, Henry held a teaching position at the Albany Academy before he moved on to Princeton in 1832. He later became the head of the Smithsonian Institution in Washington, DC. While in New York, he was involved in several geophysical projects. Like many scientists of his day, Henry held an integrated view of geophysical phenomena.[3] He believed that terrestrial magnetism, meteorology and topography were all interrelated, demonstrating the unity of science. Geomagnetism was a special interest for Henry, who saw the practical importance of adequate magnetic observations in a

---

1. Nathan Reingold, *Science American Style* (New Brunswick: Rutger's University Press, 1991), 118
2. Columbus is sometimes credited with the discovery of magnetic variation during his first voyage, although what he saw (and recognized) was the so-called "diurnal" variation. This effect was actually caused because the pole star itself does not mark exact north, and in the fifteenth century actually described a circle some 3 degrees in diameter. As a result of this motion around true north, any compass readings taken with reference to the pole star would themselves appear to vary, if the reader was assuming the pole star to be immobile. Log entries for September 13, 1492, September 17, 1492. *Log of Christopher Columbus*, Robert Fuson, translator. (Camden: International Marine Publishing Company, 1987), 42, 63.
3. Albert E. Moyer, *Joseph Henry* (Washington: Smithsonian Institution Press, 1997), 124.

country where the boundaries of estates were originally fixed and described by compass, the only instrument of convenient use for surveying in heavily forested country. Henry felt that a complete knowledge of the laws of magnetic variation would "obviate much perplexity on the part of the surveyor and much unfortunate dispute and Litegation on the part of the neighboring owners of the property."[4]

In 1825, Henry briefly joined DeWitt's magnetic survey for the state of New York.[5] The same year the academies of the state of New York instituted regular meteorological observations under Henry's superintendence.[6] In September 1830, he began a series of observations for Professor James Renwick of Columbia College to determine the magnetic intensity at Albany.[7] In 1832 Henry wrote to Benjamin Silliman that the subject of terrestrial magnetism afforded a wide field for observation and experiment. He himself had given considerable attention to the subject and hoped to be able to tour the boundary line between New York and Pennsylvania to determine the amount of the variation since the last survey, done fifty years before.[8] By 1833, Henry was able to claim that Albany was the only site in the country with a complete specification of its geomagnetic elements.[9]

It was during his surveys that Henry noticed a connection between terrestrial magnetism and the aurora borealis, then considered to be a meteorological phenomenon. Henry noted that an unusually intense aurora display had been preceded by a great increase in magnetic intensity earlier in the day. The thought occurred to him that the uncommonly brilliant appearance of the aurora might possibly be connected with the preceding magnetic disturbance. Contrary to his expectations, Henry noted a considerably lower intensity than usual during the aurora.[10] Unknowingly, Henry had confirmed an earlier observation by Christopher Hansteen of the sudden rise and fall of magnetic intensity accompanying the appearance of an aurora. Later Henry discovered that Samuel Hunter Christie had observed the same aurora across the ocean in Britain, enabling him to proudly claim that "a simultaneous disturbance of terrestrial magnetism, in

---

4. Henry to Forbes, June 7, 1836. (JHP III p.73).
5. Moyer, 121.
6. Henry to Oersted, April 27, 1841. (JHP V p.28).
7. Joseph Henry, "On a Disturbance of the Earth's Magnetism." *American Journal of Science and Arts* 22 (April, 1832), 145. Henry used "needles furnished to Prof. Renwick by Cap[tain] Sabin[e]." Henry to Silliman, March 28, 1831. *Science in Nineteenth Century America: A Documentary History*, Nathan Reingold, ed. (Chicago: University of Chicago Press, 1964), 66. Hereafter cited as Reingold.
8. Henry to Silliman, March 23, 1832. Reprinted in Reingold, 69–70.
9. Moyer, 161.
10. Henry, 1832, 146.

connexion with an aurora, has [n]ever before been noted at two places so distant from each other."[11]

Alexander Bache (1806–1867) was also interested in geophysics. The great grandson of Benjamin Franklin, Bache had attended the United States Military Academy. He was an influential member of both the Franklin Institute and the American Philosophical Society in Philadelphia, and later taught at the University of Pennsylvania and became president of Girard College. He was best known as superintendent of the United States Coastal Survey from 1843 to 1867. In 1830, Bache began making magnetic observations, inspired by the work of Gauss and Weber. He conducted a magnetic survey of Pennsylvania from 1840 to 1843. Later Bache helped to establish the first magnetic observatory in the country at Girard, which was in correspondence with the global system already in operation in the 1840s.[12]

Meteorology was another field where Americans began to make observations. The *American Journal of Science and Arts* (also known as Silliman's Journal after its editor) reported observations from Ohio in the early decades of the nineteenth century. At the same time, James Espy was involved in setting up a series of meteorological observations that he attempted to tie together into a national system. Espy and Alexander Bache set up a joint venture between the Franklin Institute and the American Philosophical Society to solicit meteorological observations in 1834. In 1837 the American Lyceum requested that individual lyceums keep weather observations and forward them to Espy. By 1839 the state of Pennsylvania had set up a statewide system of weather observations.[13] In the 1830s, the academies of the state of New York were also submitting annual meteorological reports to the regents of the university.[14]

European connections provided a considerable level of support and guidance for the American scientists. Henry and Bache were in correspondence with British geophysicists such as James Forbes and Edward Sabine.[15] As a result, the Americans could not help but be aware of the events in Britain leading up to the Magnetic Crusade. In the spring of 1837, about a year before the lobby was launched, Henry visited England and met with many prominent men in the field. He discussed geomagnetism with Sabine as well as the latter's role in building fortifications at Fort Eire during the War of 1812.[16] Henry received the praise of John Daniell, who claimed that the United States was "setting the example to the world in reference to [meteorological]

11. Ibid., 152.
12. DAB.
13. Reingold, 128.
14. Henry, 1832, 150.
15. Henry to Forbes, June 30, 1834. (SAUL msdep7 Incoming Letters 1834).
16. Henry's European Diary, April 24, 1837. (JHP III p.312).

observations."[17] Henry learned that there were already over twenty observatories established on the continent making geomagnetic observations under the direction of Gauss.[18] He also met with Christie to discuss setting up magnetic observatories to cooperate with those established by Humboldt.[19]

Eventually, both British and American scientists became part of the global geomagnetic project. However, strains between the two existed. While Brigitte Schroeder-Gudehus assures us that "scientific theory and practice are by nature international," she also admits that science could be "a matter of collective pride and envy and, therefore, a possible source of tension between co-operation and competition."[20] While geophysics necessitated a worldwide approach, it could also inspire nationalist feelings and antagonism. The discovery of the south magnetic pole could be viewed as a joint effort or a contest, a race to the finish for the glory of the home country. These tensions could be both productive and destructive, and would have to be carefully managed.

▩ ANGLO-AMERICAN SCIENCE AND STATE

The best example of science and state interaction in the antbellum period was found in the push for an American exploring expedition in the 1820s and 1830s. Such an expedition was advocated both for advances in navigation and discoveries in the South Pacific, as well as for an American contribution to the storehouse of human knowledge of the world. New Englanders were especially in favor, given the close connection of the region with mercantile and whaling interests. But the states alone could not provide the support that was needed for a full-scale expedition. By the 1820s, it was perceived that in order to succeed, any venture "must be clothed with authority from government, and the officers and men on regular pay."[21] Convincing the American government to support an exploring expedition would be a difficult task, and this example provides an interesting contrast to the push for the Magnetic Crusade that occurred in Britain during the same decade. Unlike in the American Congress, party politics did not play a great role in lobbying for the crusade in the British Parliament. The 1835 failure came under the same Whig prime minister (Melbourne) as the 1839 success, while the 1842 renewal came under a

---

17. Ibid., March 23, 1837. (JHP III p.193).
18. Ibid., August 1837. (JHP III p.494).
19. Ibid., April 21, 1837. (JHP III p.303).
20. Brigitte Schroeder-Gudehus, "Nationalism and Internationalism," *Companion to the History of Modern Science* (London: Routledge, 1990), 909–910.
21. Quoted in William Stanton, *The Great United States Exploring Expedition* (Berkeley: University of California Press, 1975), 28.

Tory prime minister (Peel). Some of the prominent supporters of the lobby (such as Lloyd) were Tories, but others (such as Northampton) were Whigs.[22] In the United States, politics played a much larger role in the negotiation between science and state. The major points of comparison are the lobby for the expedition and the interaction of science and state. In both cases, elements of the American political and social makeup influenced not only the chances for success, but also the final form of the project.

American politics in the early nineteenth century continued to enshrine the laissez-faire doctrines of the Enlightenment. Many political thinkers of the time followed Thomas Jefferson's "strict constructionist" view of the Constitution, which limited the role of government in public affairs. Strict constructionists sought to keep taxes low, government spending down and to prevent Congress from legislating in violation of the rights of the individual states. They were hesitant to spend any money on projects that were not strictly necessary and which did not generally benefit the entire nation. On the issue of whether the government could spend money for particular improvements, the answer was clear. Roads, canals and turnpikes that ran entirely within one state or were only of local benefit could not be federally subsidized. In his veto of the Cumberland Road Bill (1822), James Monroe had written that Congress did not possess the power, under the Constitution, to pass such a law.[23] This frugal approach to government limited the chances that the federal administration might become involved in scientific ventures. A government that would not spend federal money to help build intrastate roads could hardly be expected to support an expedition to the South Pacific for the benefit of whalers and merchants, much less scientists. As a result American scientists could expect little assistance from Washington, DC in the early nineteenth century.

By the 1830s, resistance to the laissez-faire ideas of the strict constructionists had developed at the national level in the form of Henry Clay's "American system." Clay advocated government spending for internal improvements, drawing a great deal of support from the

---

22. Science, however, did seem to fare slightly better overall under the Whigs. Herschel, at least, saw greater promise in the party of Earl Grey than that of the Duke of Wellington. Writing to Sabine about his concerns that Peel's government was not showing enough sympathy to their project he commented that "I cannot help regarding as exceedingly ill-omened...the prospects of Science under the new [Tory] Government. It is remiss to see so early an indication of the *good old Tory feeling* of hatred and contempt for science and its followers peeping out...[P]olitics & science have little in common & the less the better perhaps." Lloyd to Sabine, March 3, 1840. (PRO BJ3/10/24); Herschel to Sabine, April 5, 1841. (PRO BJ3/26/275–6).

23. *Annals of Congress.* 17th Cong., 1st sess., 1803. [May 1822]

developing West and the maritime states of New England. Clay's Whig party, which included Daniel Webster and John Quincy Adams of Massachusetts, believed in a more active government and was willing to allow government spending (and taxes) to increase. The Whigs were to some extent a revival of the old Federalist party, which also held loose constructionist views on the Constitution.[24]

In opposition to Clay, Andrew Jackson's Democratic party continued to hold to the Jeffersonian ideas of laissez-faire politics and a minimal state. From the 1820s to the 1840s, these two sides fought for control of the presidency and influence in Congress. Whig strength lay in New England, along the Atlantic coast and in the developing western states. Democrats were strong in the South and middle Atlantic states. Politics in the antebellum period are often analyzed with a view toward the Civil War, and thus sectional differences between the free North and the slave South are emphasized. However, party divisions were more important in determining voting patterns on most issues, even across sectional lines. Joel Sibley finds that in the period of the 1830s and 1840s "the American people were both highly politicized and highly partisan." This partisanship was in part due to the legacy of the political split in the country between supporters (Democrats) and opponents (Whigs) of Andrew Jackson's policies in the 1830s. It enabled the respective parties to hold together the majority of their members on most votes in Congress. [25] Also, in contrast to the British system that unified legislative and executive power, the American system divided power between the various branches of government. In order for a bill to become law, its supporters had to gain the consent of three different institutions, the House, Senate and presidency.[26] Opposition from any one of these could effectively block a bill.

Adams's presidency (1825–1829) marked a period of looser interpretation of the Constitution, which gave hope to scientists seeking state support. Indeed, in his inauguration, Adams had called for "laws

---

24. The period 1825–1835 actually saw an evolution of parties in opposition to Jackson's Democrats, from the administration party under Adams (1825–1829) to the National Republican party in Jackson's first term (1829–1833). By the congressional elections of 1834 and the presidential election of 1836 these anti-Jacksonians had coalesced into the Whig party. For simplicity the anti-Jacksonians are referred to as Whigs throughout this period. A strong third party was the anti-Masonic party, which was particularly influential in Pennsylvania, New York and New England. It had formed in the 1820s as a reaction against freemasonry and Jackson. In the 1830s the Anti-Masons tended to side with the Whigs, and after 1836, merged with that party.
25. Joel Sibley, *The Partisan Imperative: The Dynamics of American Politics before the Civil War* (New York: Oxford University Press, 1985), 135–140.
26. At this point in American history, the role of the Supreme Court as interpreter of the Constitution was only just beginning to be defined.

promoting the improvement of agriculture, commerce, and manufactures, the cultivation of the mechanic and of the elegant arts, the advancement of literature, and the progress of the sciences, ornamental and profound."[27] Adams later proposed a national system of astronomic observatories, or "lighthouses of the skies," for the purpose of improving geography and navigation.[28] He later became a staunch advocate of establishing a national astronomic observatory in Washington, DC.[29]

By contrast, the populist and anti-intellectual quality of Jackson's presidency (1829–1837) appeared less promising for the fortunes of science. Jackson derided the views of the previous administration that, he claimed, had overreached itself. In his veto of the Maysville Road Bill (1830), Andrew Jackson (like Monroe) upheld the principle of limited government. He clarified the position that had been tacitly accepted for some time: grants of money could be made under the principle that the works that might be aided should be of "general, not local—national, not state character." Jackson believed that any disregard of this distinction would necessarily lead to the subversion of the federal system. The crucial test for funding in the 1830s was whether a proposal was in the national interest, conductive to the benefit of the whole, or merely local, and only to the advantage of a part of the union.[30] Within the context of this struggle came the American push for an exploring expedition. The results were surprising.

THE EXPLORING EXPEDITION

The history of the United States Exploring Expedition, or "Ex Ex" as it came to be known, was long and tortuous. The American maritime community had long pushed the government to finance a project that might lead to navigational advances. In the early nineteenth century, most American sailors still had to rely upon European maps and charts. During the War of 1812, it had become embarrassingly apparent that the British possessed better charts of American coastal waters than did their former colonists. In 1828, the secretary of the navy could write, "we now navigate the ocean, and acquire our knowledge of the globe, its divisions and properties, almost entirely from the contributions of others."[31] Even by the 1830s, no

---

27. Quoted in Richard Current, *American History* (New York: Alfred Knopf, 1964), 235.
28. Stanton, 4.
29. See Marlana Portolano, "John Quincy Adams's Rhetorical Crusade for Astronomy." *Isis*, 91:3 (September 2000), 480–503.
30. U.S. *House Journal.* 1830. 21st Cong., 1st sess., 27 May.
31. Jeremiah Reynolds, *The South Sea Surveying and Exploring Expedition* (New York: Harper & Brothers, 1836), 186.

complete survey of American coastal waters had been carried out. In 1838 Congressman Henry Wise (Whig, VA) could still claim that "the British Admiralty now know more about the waters of the Chesapeake Bay than the officers of our own navy."[32] Not until the 1840s did Bache's coastal survey begin to rectify these deficiencies.

In addition, the whaling and sealing industries of New England were eager for the United States to become involved in the exploration of the South Pacific and Antarctic waters, both to open up new fisheries and to improve existing charts. The advocates of an exploring expedition argued on several grounds. First, the United States was an emerging commercial nation, rivaling Britain. The expedition would help commerce by opening new trading routes and new markets. Second, whalers and sealers were already complaining of reduced stocks in the traditional fishing grounds off of South America. New fisheries for whales as well as for seal and "sea-elephant" islands could be discovered, increasing the supplies of these animals available to fishermen. Third, many existing reefs and shoals were uncharted in the waters of the South Pacific, posing a danger to those who sailed in them. Better charts could improve navigation and thus lives could be saved. Finally, the expedition provided an opportunity for scientific discovery.

Opponents of the expedition met these points with their own. As advocates of small government, they feared the expenditure of tax dollars for anything as risky as a sea expedition into unknown waters. Many saw the proposed expedition as simply a boondoggle for New England, a federal subsidy for the whaling and sealing industries that had already overexploited existing stocks to near extinction. Additionally, many in the young republic were not inclined to become involved in international or imperial affairs, preferring to eschew European expansionist policies and to put the needs of Americans at home first.

When compared to the Magnetic Crusade, the American effort displayed both similarities and differences. The Ex Ex was similar to several proposed voyages of discovery of this time period also being sent out by various European nations. In 1835, Edward Sabine had called for a British expedition to the southern hemisphere for the purpose of geomagnetic research, a proposal that (after many changes) eventually led to the Magnetic Crusade in 1839. All these expeditions shared in the geophysical interest of the time and sought to explain

---

32. *Congressional Globe.* 25th Cong., 2nd sess., 274. [April 9, 1838] Records of debates in Congress were kept by the *Register of Debates* and later the *Congressional Globe*. Both of these were private publications that made no attempt to provide a verbatim report of what was said on the floor of the House. Instead, summaries of the debates were given, often in the third person. As a result, quotations in this paper will come from these reports, not the actual debates.

global phenomena like terrestrial magnetism through a global series of observations. But unlike the final form of the crusade, these expeditions were meant to be one-time surveys covering a lot of ground, rather than permanent observation posts throughout the world as advocated by John Herschel.

The differences in the two projects demonstrate to some extent the differences in the two nations where they were conceived. Britain, with its many institutions of science and learning, had more established men of science than the young republic, and the "aristocratic democracy" of Britain provided them with a path to appeal to the upper layers of the state directly. Personal connections between men of science and government played a large role in the success of the crusade. Joseph Henry had noted something of this sympathy when he earlier commented that the British Association "is quite as aristocratical as the government of the nation."[33] By contrast, the United States had a lower quota of scientists and a system of "frontier democracy" which forced advocates to win popular support even before the scientific community became involved.

In Britain, the lobby for the crusade involved many of the most prominent men of science of the time. They had organized to request that the government approve a project which had already been planned and in which they intended to participate. The government might provide the financial means, but the scientists would handle the details. In contrast to the extensive lobby for the Magnetic Crusade in Britain, the Ex Ex was more of a one man crusade in the United States. Jeremiah N. Reynolds, a geologist and student of John Symmes, was the main scientific proponent of the expedition.[34] Symmes had followed Halley's theory to explain magnetic variation, that the Earth was composed of several concentric spheres. Symmes developed his own idea that the Earth was open at its poles, allowing access to the inner spheres, which he believed were inhabited. In 1823, Symmes's supporters petitioned Congress for an expedition to the North Pole for the purpose of locating such an entrance, both for the sake of geographical discovery and to open up trade with the interior inhabitants! While the House rejected the petition by a vote of 56 to 46, later polar expeditions continued to be associated with Symmes's unorthodox ideas.[35]

Although divorcing himself from Symmes's "hollow earth" theories, Reynolds still supported his teacher's desire for a polar expedition— not to find the supposed inhabitants of the inner spheres, but to locate a southern continent and make various geophysical observations along the way. Reynolds was involved in the proposed expedition for over a

---

33. Henry to Bache, August 9, 1838. (JHP IV p.103).
34. Nathaniel Philbrick, *Sea of Glory* (New York: Viking, 2003), 20.
35. *Annals of Congress.* 17th Cong., 2nd sess., 698–699. [January 1823]

decade, suffering the vicissitudes of American politics as his hopes for an expedition were first buoyed then dashed by successive administrations. Unlike Herschel and Sabine, Reynolds did not have a well-organized lobby to back him up. While he had the support of many scientists, there was nothing in the United States that approached the scope of the magnetic lobby in Britain. Both the limited numbers of American scientists and the lack of any centralized organization to coordinate them, such as the Royal Society or the British Association, no doubt contributed to this lack of support.

Reynolds also lacked the political connections of someone like Herschel in Britain, but social connections among gentlemen would not have been enough to accomplish the same results in the United States even if they had existed. The American political system did not allow such an appeal. In Britain, approval for the Magnetic Crusade came from the ministry with no reference to Parliament and was funded by a special grant from the treasury. The lobby for the crusade worked with the very people who would implement its plans from the beginning. In the United States, this approach could not have worked. The division of powers in the American government served to prevent a major project from being initiated by the executive without congressional approval. A personal relationship with the president or the secretary of the navy did not guarantee approval by the House and Senate, which controlled funding. And the executive branch could not always be trusted to carry out Congress's wishes even after they were passed into law, as Reynolds later discovered. In order to get a proposal through Congress the expedition had to gain public support and be able to navigate the various interest groups in Congress. The fate of the first proposal for an exploring expedition in 1828 showed just how many things could go wrong.

On May 19, 1828, John Reed (Whig, MA) introduced a resolution to send a vessel to the Pacific Ocean and the South Seas, "to examine the coasts, islands, harbors, shoals and reefs, in those seas, and to ascertain their true situation and description."[36] This resolution came out of months of research that had been conducted by the House naval affairs committee. Reynolds had been asked to contribute a report, speaking on the issue of commerce in the South Seas. He noted that industries dependent upon the whale, seal, sea otter, sandalwood, feather and "sea elephant tooth" trades would benefit from the expedition.[37] The hunting of whale and seal, heretofore carried on with so much vigor, had produced the natural consequence of reducing their numbers.[38] Also reporting to the committee was the Secretary of the Navy Samuel Southard, a friend of Reynolds. Southard endorsed the idea of the

---

36. U. S. *House Journal.* 1828. 20th Cong., 1st sess., 19 May.
37. Reynolds, 174.
38. Ibid., 179.

expedition, on the grounds of commerce, navigation and science. The increasing American commerce in the region needed the protection of the government, which would be greatly extended by such an expedition. Additionally, they could add something to the common stock of geographical and scientific knowledge.[39] The final committee report stated that

> The dangers to which an immense amount of property is exposed, as well as the hazard of human life, for the want of knowledge, by more accurate surveys, of the regions to which our commerce is extending, and the probable new sources of wealth which may be opened and secured to us, seem to your committee not only to justify, but to demand the appropriation recommended; they therefore report a bill for the purpose.[40]

The result of the committee's research was a bill (H.R. 240) "to provide for an exploring expedition to the Pacific Ocean and South Seas," which had been sent to the committee of the whole House on March 25. Despite the favorable reports, the chairman of the naval affairs committee, Michael Hoffman (Democrat, NY), had not brought the bill to the floor of the House because he personally opposed it. Reed's resolution was an attempt to force the issue. Hoffman objected both to the resolution and to the manner of its introduction. Reed replied that the resolution had the support of every individual of the committee except Hoffman himself, and that he was presenting it on their behalf. Members on the other side of the aisle supported Reed as well. Churchill Cambreleng (Democrat, NY) declared that the Pacific Ocean should become an American Sea, and expressed a strong hope that the resolution would pass. Hoffman's motion to table the resolution failed. The bill, along with $50,000 in funding, was approved without further debate two days later at the end of the session.[41]

The expedition had its first victory, but the debate was not yet over. President John Quincy Adams, who had supported the resolution, confided to Reynolds that no bill passed in that session of Congress had given him more pleasure than Reed's resolution.[42] His administration began making plans to carry out the expedition, even before the Senate took up the matter. The vessel *Peacock* had to be rebuilt, and two other ships employed. The rush was partially due to partisan politics. 1828 saw a rematch between Adams and Jackson for the presidency, and Adams feared that a Democratic victory would not bode well for the expedition. Indeed, Jackson and the Democrats were victorious in 1828, winning the election by attacking the "elitist" arrogance of the Whigs and Adams's supposed abuses of presidential power. However,

---

39. Ibid., 186.
40. Ibid., 184.
41. U. S. *House Journal*. 1828. 20th Cong., 1st sess., 21 May.
42. Stanton, 17.

the Democrats did not have to wait for Jackson to enter the White House to scuttle the expedition.

The House bill met its fate in the Senate in February 1829. Leading the opposition was Robert Hayne (Democrat, SC) a strict construction- ist Jacksonian who was determined to prevent the expedition. He was also eager to frustrate the plans of the Adams administration in its last days.[43] Hayne assailed the secretary of the navy for beginning prepara- tions for the expedition before the bill had passed both houses of Congress, and for using appropriations made for general naval purposes to do so.[44] Without Senate approval, the administration had begun se- lecting naval officers and a scientific corps including an astronomer, a naturalist, draughtsman, and surveyors to accompany the expedition. Hayne saw in these acts a threat to the balance of powers prescribed by the Constitution.

Thus Hayne was attacking the Adams administration as much as the bill, though he had little sympathy for the scientific aims of the expedition anyway. He felt that science had no place in a naval expedi- tion. He charged that as the *Peacock* expedition was presently orga- nized, the officers of the navy were to be mere navigators. Hayne maintained that the naval officers should be at the head of the mission. The scientific corps should only be their agents and instruments. To the navy should belong the glory of the enterprise, if any glory was to be acquired in it.[45] Faced with such opposition, the bill could not sur- vive the overwhelmingly Democratic Senate. On the final vote, the bill was defeated 13–27. Twelve Whigs (seven from New England) joined by one Democrat (from New Hampshire) supported the expedi- tion.[46] Against it were twenty-one Democrats (sixteen of them from the South), and six Whigs. Support for the expedition lay with the Whigs (67% in favor) and the New England senators (73% in favor). Opposition came from the South (95% opposed), while the West split against the expedition. Only two Senators from inland states voted for the bill as opposed to nine against, implying that the expedition was still seen largely as benefiting the coastal states. Because it failed to develop political support outside of the Whig party or among the repre- sentatives of the inland states or conservative southerners, the bill could not pass.

The failure of the 1828 *Peacock* expedition demonstrated the diffi- culties that any proposal would face. All sides felt let down. Sailors complained that the scientific baggage added by the administration

---

43. At this point, the new president was inaugurated in March.
44. *Register of Debates.* 20th Cong. 2nd sess., 51. [February 5, 1829]
45. Ibid., 52. [February 5, 1829]
46. For the purposes of this analysis I take New England to be the six states of the northeast, the South to be any slave state and the West to be any non- slave state not on the coast.

had sunk their expedition; while scientists such as botanist John Torrey complained that the expedition "was destined, not for discovery and for scientific purposes—but to catch seals!"[47] However, supporters of the proposal had learned several valuable lessons. First, in order to succeed the expedition had to be perceived as being in the national interest. The political climate of the time doomed any bill that seemed to be sectional or which benefited only one part of the country. Secondly, the naval and commercial benefits had to be stressed over the scientific ones. The anti-intellectualism of the egalitarian Democrats would not quickly embrace a "visionary" scientific expedition. Finally, the administration that could pass such a proposal was not that of the departing Whig Adams, but that of the incoming president, Andrew Jackson. Adams and the Whigs were too encumbered by their New England roots and apparent bias for the expedition for it to have a real chance of passage. Ironically, the proposal was more successful under a Democratic administration which appeared disinterested and which could not be so easily assailed in Congress by strict constructionists.[48]

Supporters of the expedition tried again six years later. In December 1834, Dutee Pearce (Anti-Mason, RI) presented a petition from Reynolds for an expedition to survey the islands and reefs of the Pacific and its coast. Pearce argued that the annual loss of property upon the islands and reefs not laid down upon any chart was equivalent to the expense of the expedition and survey requested.[49] In February 1835, Pearce reported a bill (H.R. 719) "to provide for an expedition to the Pacific Ocean and the South Seas" out of the committee on commerce. He made no delay in getting to the national character of the proposed expedition: "The intercourse between the different parts of the nation and the islands and countries of the Pacific has become a matter of public interest, and deserv[es] the protecting care of the national legislature."[50] In March 1836, the Senate committee on naval affairs, now chaired by Samuel Southard (Whig, NJ) recommended an exploring expedition to the Pacific Ocean and South Seas.[51] The House committee on naval affairs followed suit the same month with its own report concentrating on the commercial aspects of the expedition.[52] The report also emphasized national honor, and the duty that the

---

47. Quoted in Stanton, 27.
48. Strict-constructionist Democrats ironically launched most of the important American expeditions of this period. William Goetzmann, *New Lands, New Men* (New York: Viking, 1986), 270.
49. *Register of Debates.* 23rd Cong., 2nd sess., 778. [December 10, 1834] The register records the petition as being from John N. Reynolds but this must be a mistake for Jeremiah N. Reynolds.
50. Reynolds, 243.
51. Philbrick, 30.
52. Reynolds, 265–270.

government and the nation owed to the common cause of all civilized nations, the extension of useful knowledge of the globe.[53] It was also the duty of Congress to extend, secure and protect every portion of American commerce.[54]

In April 1836, Pearce arranged for Reynolds to address the House of Representatives concerning the proposed expedition. As was to be expected, Reynolds's appeal featured nationalism, honor and safety for commerce. Reynolds lamented the fate of the *Peacock* expedition in 1828, believing that its failure should have been "a subject of regret to every enlightened mind in the least acquainted with the subject, without reference to party, profession or sectional feelings."[55] Reynolds's goal remained the same: a naval voyage of discovery to be fitted out with every scientific instrument, at the public expense, for the sole purpose of increasing the knowledge of the Pacific, where American commerce was now carried.[56] Challenging the idea that the expedition would only benefit fishermen, Reynolds declared that as the fisheries reach the interest of every class of citizens in the country, their prosperity was to the advantage of the whole people. Reynolds also spoke to the duty of the government to afford every facility to the merchants in these commercial enterprises, and to furnish them with adequate protection.[57] Only after he made his case for navigation and commerce did Reynolds mention science as an objective, comparing the expedition to Jefferson sending Lewis and Clark out into Louisiana.[58] Reynolds's address marked the culmination of the latest push for the Ex Ex.

Fortunately for supporters of the expedition, intervening elections had changed the composition of the Senate, increasing Whig representation and making that body more sympathetic. By May 1836, the Senate had already passed a Naval Appropriations bill containing an amendment authorizing the president to dispatch an exploring expedition to the Pacific Ocean and the South Seas.[59] Thus the final debate this time was in the House. In his treatment of the expedition, Nathaniel Philbrick grossly oversimplifies the route which the bill for the expedition took through the House, merely claiming that Reynolds's "stirring and patriotic call to science resonated with Congress." The heavy resistance the bill would face he dismisses as "a slight ripple of protest."[60] In fact, the Ex Ex confronted serious obstacles in the House which had to be overcome before the bill could become law. Opponents of the expedition tried on three

---

53. Ibid., 268.
54. Ibid., 267.
55. Ibid., 30.
56. Ibid., 25.
57. Ibid., 45–46.
58. Ibid., 75.
59. *Congressional Globe.* 24th Cong., 1st sess., 418. [May 5, 1836]
60. Philbrick, 31.

successive occasions to derail the proposal, and only by making serious compromises and changing the expedition's focus from science to discovery could its supporters succeed in passing the bill. The history of the voyage is well known and has been covered in many works. I am concerned here not with the actual expedition, but with the political activities that brought it about in the first place. It is interesting to observe the process through which the legislation gained the approval of the American Congress and to note the ways that process helped to shape the final form of the expedition itself.

Attempts to alter the bill began at once. In the House, the committee on naval affairs recommended an amendment to the Senate bill. Sponsored by Leonard Jarvis (Democrat, ME) this amendment changed the authorization to allow the president to dispatch the expedition only if in his opinion the public interest required it. The issue of separation of powers had come up again. Some Democrats did not like the idea of Congress directing the president to send out an expedition, which they believed was implied in the Senate bill. Rather, they preferred to leave the choice up to the chief executive. However, the constitutional question reflected another argument. The debate over the constitutionality of the bill soon broke down into a debate over the expedition itself. Supporters of the expedition were willing to vote for the bill, amendment or not. Opponents voted against the amendment as a way of voting against the expedition. They believed that without the amendment, the overall bill would fail when brought to a final vote. These maneuvers could be seen in the events of the next few days.

Upon taking up the bill on May 6, debate soon shifted as opponents of the expedition began to attack it directly. Albert Hawes (Democrat, KY) referred to Reynolds's address before the House, which he had opposed. According to Hawes, Reynolds

> knew that a proposition would be brought before Congress to expend a large sum of money to further [his] object. It appeared to [Hawes] that they had arrived at a time that, no matter what chimerical opinions were taken up by any gentleman, if he could only get an opportunity to address the Representatives of the people they would enter into all the chimerical notions of the gentleman who might address them.[61]

Hawes, a good Jacksonian, believed that the contents of the Treasury had been wrested from the hands of the people of the United States, and was not willing that it should be expended in every "chimerical" notion that entered the head of any individual. He compared the expedition to Adams's famous proposal for "lighthouses of the skies" which Hawes saw as "a ladder to happy places" and upon which he blamed Adams's loss in 1828. Hawes derided the expedition as an

---

61. *Congressional Globe.* 24th Cong., 1st sess., 423. [May 5, 1836]

attempt to take the vessels and seamen of the United States and send them to the South Seas exposing them to all the diseases, hurricanes and mishaps of that climate. And for what gain? "For nothing on the face of the earth," Hawes declared.

John Patton (Democrat, VA), representing the isolationist and anti-imperial tendencies of his party, was also uneasy. Was the purpose of the bill an expedition for the improvement of science, or the discovery of unknown regions to be taken possession of and colonized by America? While he opposed the expedition, Patton also opposed Jarvis's amendment for shifting the legislative functions upon the shoulders of the executive departments. A strict constructionist, Patton lamented that "to all practical purposes of legislation in this country they were as perfectly unrestrained by the Constitution as if that measure had been thrown into the fire the day after it went into operation."[62]

Support for the amendment (and the expedition) came from Whigs and some northern Democrats, especially those representing coastal regions. Samuel Vinton (Whig, OH) cited the vast commerce the United States had in the South Pacific without any maps or charts, in seas more perilous than any other. He had no problem leaving the final decision to the president. If Jackson should be satisfied that there was a total want of those maps and charts then let him order the expedition be fitted out. Vinton believed that since America had done less than any other part of the globe for science, "we should also contribute our mite."[63]

When debate resumed on May 9, the main subject was still the expedition, rather than the amendment. John Reed (Whig, MA), the sponsor of the 1828 resolution, saw it as truly a national object.[64] He knew that the supporters of the expedition were favoring the bill with or without the amendment, and he did not oppose the amendment lest the bill be defeated in the final vote. While acknowledging that a large portion of the nation's whaling industry was in his district, he was pleased that some inland representatives residing far from the ocean were also in support of the proposed expedition.[65] He believed that all parts of the country had a common interest in seeing the expedition approved.[66]

---

62. Ibid.
63. Ibid., 424.
64. Ibid., 440
65. Reed's eleventh congressional district included Nantucket, Dukes and Barnstable counties. The 1840 census listed 7,474 people employed in seafaring there, representing about one-sixth of the total population of 45,518 in those counties. *Sixth Census or Enumeration of the Inhabitants of the United States* (Washington, DC: Blair & Rives, 1841), 10–11.
66. *Appendix to the Congressional Globe.* 1836. 24th Cong., 1st sess., 569–573.

Thomas Hamer (Democrat, OH) picked up on Reynolds's reference to Lewis and Clark, claiming that their expedition was exactly like the one now proposed. He also saw the expedition as a matter of national import that deserved congressional support just like navyyards, lighthouses, forts, arsenals, dockyards and harbors. Hamer also found reason for inland Western states to back the project.

> Some gentlemen appeared to consider this an Eastern measure. It was not so. The West had a deep interest in it. It was well known to all who resided in the great grain-growing States of the interior, that our principal difficulty was to find a market for the surplus production of our fertile soil.[67]

Hamer claimed that every nation in Europe had sent out similar exploring expeditions, to which Hawes replied that he was not impressed. "They had their dukes, and their marquises, and lords, but that was no reason why we should."[68] Hawes saw no practical benefits resulting from the expedition and suggested that if the bill passed, "they should next have a proposition for a voyage of discovery to the moon."

Stephen Phillips (Whig, MA) denied that the expedition was "visionary" and was satisfied that it would diminish the danger of South Sea voyages. He also believed that important commercial advantages might grow out of such an expedition. Finally, Philemon Dickerson (Democrat, NJ) concluded the debate by pointing out that the naval services bill had been delayed for weeks in discussing a measure for an exploring expedition that, if carried into effect, it was doubtful whether any benefit would result. (This is not the last we hear of the Dickersons or their doubts.) Churchill Cambreleng (Democrat, NY), who had supported Reed in 1828, now called for a vote. On the question of Jarvis's amendment to the Senate bill, the amendment was agreed to 92–68.[69]

A close analysis of this vote is of interest, since support for the amendment tended to reflect a positive opinion toward the expedition. Party, location and sectional affiliations all influenced the way in which representatives voted. One immediate factor in deciding whether to vote for the expedition was party affiliation. The expedition enjoyed the overwhelming support of the Whigs (71% in favor), while Democrats split against it (52% opposed). Local conditions were also a major influence. Representatives from coastal districts and New England districts (83% in favor) tended to favor the expedition over

---

67. Ibid., 337–340.
68. *Congressional Globe.* 24th Cong., 1st sess., 440. [May 9, 1836]
69. Ibid., 441; U. S. *House Journal.* 1836. 24th Cong., 1st sess., 9 May. In favor were thirty-nine Whigs, forty-seven Democrats, and six Anti-Masons Opposed were sixteen Whigs, fifty-one Democrats, and one Anti-Mason.

inland ones. Finally, sectional differences contributed as well. 78% of southerners voted against the measure, including all of the Whigs who opposed it, although local factors could still win out over party and section: southern Whigs (as well as northern Democrats) from coastal districts tended to favor the expedition.

In the end though, party loyalty was the most important factor. Here the Whig's traditional strength in the West helped immensely. The deciding votes for the amendment came from a group of inland Whigs representing the states of Ohio, Indiana, Kentucky and Tennessee.[70] Isolated from the coast, these representatives had no immediate reason to support the expedition, as it offered no direct benefit to their constituents. Yet the majority of these Whigs (twelve out of twenty-one) voted in favor of the expedition anyway, providing the margin of victory. Whig party discipline was clearly an important factor in deciding their votes.

Although the matter seemed to be settled, on May 10 the House took up the proposition again. On sending the House bill to the Senate, that body had struck out the Jarvis amendment. Now the House either had to approve the original Senate bill, or else the entire expedition would fail. Again the lines broke down between supporters of the expedition and its opponents. Most of those voting on May 10 kept their original vote. The same representatives who voted "yea" for the Jarvis amendment on May 9 now voted "yea" to go with the original Senate version. For these supporters of the expedition, the amendment had always been secondary. John Quincy Adams summarized their position well when he remarked that "the bill [had] the same effect, whether the words struck out by the Senate were retained or not." The final vote came up 79 to 65 in favor of the Senate version and the expedition.[71] Fewer than 10% of those who voted both days had changed their position. For this vote 70% of Whigs again stood by the expedition while 52% of Democrats still opposed it. Again, the inland Whigs provided the margin of victory, splitting seven to seven. While local and sectional interests played a role in the outcome of the vote, in the end party discipline enabled the Ex Ex to prevail. The Ex Ex won because the Whig party was better able to keep its members in line behind the proposal than the Democrats were in opposition to it, enabling the expedition to win out despite an overall Democratic majority in the House. The Ex Ex finally had its vote of confidence.

---

70. Ohio's delegation voted heavily for the amendment regardless of party, perhaps influenced by Reynolds, who was from Ohio.
71. Ibid., 445; U. S. *House Journal.* 1836. 24th Cong., 1st sess., 10 May. In favor were thirty-one Whigs, forty-five Democrats, two Anti-Masons and one independent. Opposed were thirteen Whigs, forty-nine Democrats, two Anti-Masons and one independent.

However, to win this political victory, supporters of the expedition had been forced to recast it. Sectional benefits for New England and scientific objectives were downplayed. Only by convincing Congress of its national importance, economic benefits and navigational prospects had supporters been able to gather the broad support needed to counter the majority Democrat inertia against state expenditure.[72] In the process, though, the expedition had lost some of its scientific content, a trend which would continue into the implementation phase.

🏵 PROBLEMS WITH IMPLEMENTATION

The passage of the naval appropriations bill with the amendment for the exploring expedition had been good news for Jeremiah Reynolds. Even better news soon came: the Ex Ex had caught the nationalist fancy of Andrew Jackson himself, who showed no indications of resisting the will of Congress even if it did mean the expenditure of $100,000. Jackson wrote to his Secretary of the Navy, Mahlon Dickerson, that summer stating that he was "feeling a lively interest in the exploring Expedition directed by Congress." It was Jackson's desire that measures be taken to prepare and complete the outfitting of the ships. Jackson appointed Commodore Thomas Jones to begin recruiting the expedition crew. The president also felt strongly that Reynolds ought to go with the expedition.[73] That summer the *Washington Globe* announced that

> we learn that the President has given orders to have the exploring vessels fitted out with the least possible delay. The appropriation made by Congress was ample to ensure all the great objects contemplated by the expedition, and the executive is determined that nothing shall be wanting to render the expedition, in every aspect, worthy the character and great commercial resources of the country.[74]

It was at this point, only after the final approval by Congress, that the American scientific community became seriously involved in the Ex Ex. Unlike in Britain, where the British Association had provided a point of contact for the members of the magnetic lobby, in the United States support for the Ex Ex had been more diffuse. Since science had proved a liability during the 1828 debate over the *Peacock* expedition, it had been relegated to a secondary position in the 1836 debate. With the mission approved, it was now possible for the scientific aspect of

---

72. Goetzmann suggests that it was "Jackson's forces [that] pushed the exploring expedition bill through Congress," an assertion which is disproved by the relative lack of support from Democrats (48% in favor) when compared to the anti-Jacksonian Whigs (71% in favor). Goetzmann, 271.
73. Jackson to Dickerson, July 9, 1836. Reprinted in Reingold, 110.
74. Reynolds, 299.

the expedition to be fitted out. Accordingly, prominent scientists around the country were tapped for assistance in planning. Botanist Asa Gray was appointed to the expedition. Torrey wrote to Henry in September, asking if he had heard of the great national voyage of discovery and asking if Stephen Alexander was available to participate as astronomer.[75] Torrey also recommended Henry to Secretary Dickerson as an advisor for meteorology and magnetism.[76] Dickerson thus wrote to Henry requesting his aid to propose a set of inquiries, to be made by the scientific corps attached to the "south sea exploring expedition."[77] Henry also received a request from the Naval Lyceum for his assistance in suggesting topics for the expedition in geology, mineralogy, and especially in meteorology and magnetism.[78]

Reynolds was enthusiastic about Jackson's support and the incorporation of science into the expedition. He wrote to the president that the success of the expedition rested primarily upon the scientific corps. Reynolds recommended various heads of scientific investigation for the expedition, including geology, mineralogy, botany, zoology, meteorology, magnetism and electricity, astronomy and philology. Reynolds made it clear that he believed that the expedition was still a scientific one, and that it should not be rushed. "The time of sailing is a matter of much less moment. It should be immediately on the completion of the preparations, not an hour before."[79]

Unfortunately for Reynolds and company, all was not fair sailing. Despite the support of Congress and Jackson, the Ex Ex now had to navigate through the bureaucracy of the Department of the Navy. Here was another contrast to the British experience in 1838. The British magnetic lobby was able to work with the admiralty almost from the beginning of their effort. Herschel and Sabine were connected with naval men such as Beaufort, Ross and Minto. As a result, in Britain, the naval community was onboard from the beginning. Not so in the United States. Congressional approval and presidential support was not enough to move the navy to quick action. Additionally the current Secretary of the Navy, Mahlon Dickerson, was the brother of Representative Philemon Dickerson, who had vocally opposed the whole proposition in the House the previous spring.[80] Mahlon was as unenthusiastic as his brother about the Ex Ex and was also an old rival of Reynolds's friend, former Secretary of the Navy Samuel Southard.[81] Despite Reynolds's plea that no professional pique or petty jealousies

---

75. Torrey to Henry, September 26, 1836. (JHP III p.105).
76. Torrey to Dickerson, December 3, 1836. (JHP III p.127).
77. Dickerson to Henry, December 15, 1836. (JHP III p.131).
78. U.S. Naval Lyceum to Henry, September 27, 1836. (JHP III p.109).
79. Reynolds to Jackson, November 16, 1836. Reprinted in Reingold, 112.
80. *Congressional Globe.* 24th Cong., 1st sess., 441. [May 9, 1836]
81. Philbrick, 32.

should be allowed to defeat the object, Secretary Dickerson threw every obstacle he could into the path of Commodore Jones and his planning of the expedition.[82]

Early on, Asa Gray detected a strong attempt to break the expedition.[83] Dickerson tried to cut down the number of ships allotted and insisted that the scientific corps had to be composed primarily of naval personnel. He also tried to cut the salaries of the scientists who were chosen for the expedition. It was at this point that Charles Wilkes became involved with the expedition. Wilkes was a naval officer and the brother-in-law of James Renwick, a prominent American scientist and professor at Columbia College, who introduced Wilkes to the physical sciences and surveying.[84] Wilkes's background encouraged him to seek a naval appointment on a scientific voyage, and he had met with Reynolds at the time of the 1828 *Peacock* effort, hoping to secure a place with the expedition. Now he became an important figure in the Ex Ex. Put in charge of acquiring charts and instruments for the expedition, in July 1836 Wilkes wrote to the acting secretary of the navy that the necessary equipment was unobtainable in the United States, and would have to be sought in Europe at a cost of $15,000.[85] Thus Dickerson dispatched Wilkes to Europe on a purchasing expedition.

While in Europe (August to November 1836), Wilkes met a number of the British geomagnetic scientists, including Sabine and Ross, who were then in the process of recommending their own venture. Wilkes also visited the continent, where he found that the subject of magnetism interested them "more at the present moment than any other investigation."[86] Sabine met with Wilkes, and even recommended to him a "statical needle" for the trip.[87] They also discussed pendula experiments that could be carried out as well as other preparations for the Ex Ex.[88] Wilkes was in correspondence with Ross, asking him for advice on correcting the compass for magnetic interference,[89] and with Beaufort, who he encountered at the admiralty while borrowing charts.[90] Wilkes also attracted the attention of Charles Darwin, who hoped to meet with the American and discuss his upcoming voyage before Wilkes left London.[91] Wilkes was enthusiastic about the support

82. Reynolds, 98.
83. Quoted in Stanton, 50.
84. Philbrick, 22.
85. Wilkes to Boyle, July 18, 1836. (DU Wilkes).
86. Wilkes to Sabine, October 26, 1836. (APS HS #1:6).
87. Sabine to Lloyd, October 8, [1836]. (RS Te #18).
88. Wilkes to Sabine, November 28, 1836. (APS HS #1:6); Sabine to Hansteen, November 14, 1836. (ITA).
89. Wilkes to Ross, November 22, 1836. (APS HS #1:6).
90. Wilkes to Dickerson (draft), November 1836. (DU Wilkes); Beaufort to Wilkes, June 10, 1837. (DU Wilkes)
91. Darwin to Wilkes, Oct/Nov 1836. (APS B:D25.192).

and well wishing that he received from European governments and scientific societies.[92]

While Wilkes shopped in Europe, the expedition had been put on hold. When Wilkes finally returned it was discovered that he had forgotten to acquire several key instruments (including microscopes) and later mislaid some of the chronometers intended for the expedition. Only by constant appeals to President Jackson was Commodore Jones able to move the expedition forward at all.[93] There was also a controversy over the appointments of some "political" scientists, men who received their positions because of their connections to the Jackson administration rather than their abilities. Asa Gray feared that the expedition might be so marred by improper appointments as to render it unadvisable for him to continue his involvement.[94]

Congress was also getting anxious. In February 1837, a resolution was introduced into the House demanding that the secretary of the navy be directed to communicate to the House the number of vessels designed and fitted out for the exploring expedition with all relevant orders to and from the department.[95] Unfortunately, Jackson had left office in the spring of 1837 and his successor, Martin van Buren, was not as informed about the expedition, allowing more opportunities for delay. John Torrey's assessment in November that there was every reason for believing that the Ex Ex would sail about the end of the month proved overly optimistic.[96] By that time, Jones was tired of the constant delays and resigned his command. Reynolds later chastised "the tardy action of two secretaries—the one imbecile, and both vindictive—in fitting out an expedition authorized by repeated acts of Congress, [which] went far to weary and disgust the public mind."[97]

As if the administrative quagmire was not enough, the delays in implementing the expedition allowed time for new challenges in Congress, fostered by Dickerson. Opponents again attacked the expedition on the grounds of American interests above international ones and practical naval needs above speculative science. Henry Wise (Whig, VA) moved a resolution to convert the exploring expedition into a coastal squadron for surveying the coasts and protecting American vessels.[98] William Montgomery (Democrat, NC) also favored keeping the expedition in American waters rather than sending an expedition

---

92. Wilkes to Dickerson, October 13, 1836. (DU Wilkes).

93. Stanton, 35–56.

94. Gray to Hooker, October 10, 1836. Reprinted in *Letters of Asa Gray*, Jane Gray, editor (New York: Houghton, 1893), vol. 1, 61. Gray resigned from the expedition on July 10, 1838, shortly before it departed. Gray to Trowbridge, July 18, 1838. Ibid., 66.

95. *Congressional Globe.* 24th Cong., 2nd sess., 151. [February 3, 1837]

96. Torrey to Henry, November 1, 1837. (JHP III p.517).

97. Reynolds, 1841, iv.

98. *Congressional Globe.* 25th Cong., 2nd sess., 273. [April 9, 1838]

to "Symme's cylindrical [sic] world"![99] Supporters of the expedition
struck back. John Reed (Whig, MA) denounced this attempt to destroy
the Ex Ex and Charles Mercer (Whig, VA) argued it to be the duty of the
country, being the second commercial country in the world, to add its
quota to the cause of science.[100] Wise's attempt to convert the Ex Ex
into a coast guard was beaten back on April 4, but the greater challenge
was still to be met.[101] The expedition's opponents in Congress received
one more chance to sink it by cutting off its funding. On April 10,
Churchill Cambreleng (Democrat, NY) brought a motion to the floor
of the House to strike out the appropriation altogether, thus discon-
tinuing the expedition. This motion brought the House to a direct vote
on whether it would or would not continue the Ex Ex.[102]

Science again threatened to become a liability to the proposal. The
new scientific content that had been added to the expedition did not
find favor with all members of the House. In order to salvage the expedition, its supporters would have to compromise on its scientific aims
in order to pick up necessary votes from representatives who were reluctant to continue funding a scientific voyage during a period of economic uncertainty.[103] Isaac Crary (Democrat, MI) represented a group
of Democrats who were ready to approve the expedition without its
explicitly scientific content, accepting the need for a geographical survey while opposing the attached biological, ethological, geological and
botanical surveys.[104] Crary (rather unfairly) laid the blame for the expedition's delay on Commodore Jones, who he claimed had changed the
whole design and intention of Congress from an exploring to a scientific voyage, when the latter object should have been a secondary
one.[105] Crary claimed that the squadron would have been at sea long
ago had Jones followed the original intent of the House. But instead,
the goals of the expedition had been enlarged to include excessive scientific aims. As a result, Crary charged that Jones had loaded down the
naval expedition with unnecessary baggage:

> He [Jones] must have a body guard of scientific men from all "of the
> most celebrated institutions of the country" to catch birds and flies,

99. Ibid., 295.
100. *Congressional Globe.* 25th Cong., 2nd sess., 274.
101. Ibid., 280.
102. Cambreleng himself personally supported the expedition, but had agreed
to bring the motion against it to the floor on behalf of an absent col-
league. Ibid., 295.
103. The Panic of 1837, one of the most severe economic depressions of the
nineteenth century, had begun shortly after van Buren had taken office.
104. Crary certainly had no problem with asking for state funding in aid of
navigation: in February 1838, he had petitioned the House for an appropriation to erect several lighthouses along waterways in his home state of
Michigan. U. S. *House Journal*, 1838. 25th Cong., 2nd sess., 19 February.
105. *Congressional Globe.* 25th Cong., 2nd sess., 296.

toads and fishes. This was all very pretty...It was very well to take advantage of so favorable an opportunity to augment the stores of science,
but that object should have been only incidental and subordinate to the
great purpose of the squadron.[106]

Echoing Hayne's attack on the original bill for the Ex Ex, Crary
charged that Jones had unacceptably subordinated naval interests to
scientific ones. Despite his original reluctance to support the expedition, Crary was still prepared to back the venture given the time and
effort that had already gone into its planning. He felt that all of the
reasons the Ex Ex had been approved in 1836 were still valid in 1838.
However, Crary (among others) felt that Congress had no reason to feel
confident in the success of the expedition "as it was [then] organized;
and he was glad that an opportunity was afforded of placing it upon a
proper footing."[107] Crary's message was clear: the Ex Ex could survive,
but only as a surveying mission, stripped of the bulk of its scientific
character.

On one final vote lay the fate of the Ex Ex. The proposal was to cut
off funding for the expedition if it had not launched by May 1, 1838
(within three weeks, an impossible goal). With the help of Crary and
other Democrats, supporters of the expedition were able to hold together a coalition of representatives. The motion to derail the expedition was defeated 57–91.[108] While the deciding votes came from
Democrats like Crary who supported the expedition now that its focus
was no longer on science, Whigs were more lukewarm toward the revised goals of the expedition. This change in backing was evident in
the vote: for the first time in all the debates on the Ex Ex, a larger pro-
portion of Democrats backed the expedition (61% in favor) than did
Whigs (58% in favor).

This change was also apparent when comparing the original 1836
vote on final passage with the 1838 vote to continue funding. Most of
the representatives who voted both times kept their same position,
either for or against the expedition. However, eleven representatives
who were present for both votes switched their position, from opposition to the expedition in 1836 to support for it in 1838. 73% of these
votes were Democrats, mostly from the South. To appeal to representatives such as Crary, and to pick up conservative Democratic support,
the scientific aims of the expedition had been curtailed. The Ex Ex had
survived its last trial, but in the process a radical shift in its focus had
taken place.

---

106. Ibid., *Appendix*, 263.
107. Ibid., 264.
108. U. S. *House Journal*. 1838. 25th Cong., 2nd sess., 10 April. In favor of can-
celing funding for the expedition were thirty-two Whigs and twenty-five
Democrats. Opposed were forty-five Whigs, forty Democrats, five Anti-
masons and one independent.

Following Jones's resignation the previous year, Charles Wilkes had been appointed commander of the expedition. He began putting the Ex Ex back on a footing to launch in the fall of 1838. But the partisan bickering over science had changed the goals of the expedition, allowing Wilkes to continue weakening its scientific components. One of Wilkes's first moves was to cut down on the number of scientists to be included in the expedition.[109] Indeed, there is some evidence that Wilkes was chosen to command the Ex Ex not because of his "preeminent scientific attainments, or his particular fitness, but because he had pledged himself to dismiss a part of the scientific corps."[110] Wilkes believed that naval officers were fully competent to handle all of the scientific objects of this expedition: astronomy, surveying, hydrography, geography, geodesy, magnetism, meteorology and physics. The remainder of the scientific duties he originally planned to assign to the expedition's medical corps.[111] Although this reassignment proved impractical, he still reduced the scientific personnel from twenty-three to nine.[112]

The primary mission of the Wilkes expedition was to be exploration and surveying; science had become a secondary goal.[113] The final plan of the Ex Ex echoed Crary's reservations about its scientific content from the congressional debate. The instructions for the expedition stated that the principle objects of the expedition were to be the promotion of the "great interests of commerce and navigation." Only on occasions when not incompatible with those greater goals was it to "extend the bounds of science, and promote the acquisition of knowledge."[114]

Curiously, the Ex Ex has gone down in history as the "Wilkes Expedition," despite Wilkes's late entry into the matter. Leaving Hampton Roads, Virginia, in August 1838 (just as the magnetic lobby was getting underway in Britain), it spent the next three and a half years exploring the Pacific and Antarctic. In addition to exploration, the expedition brought back thousands of samples of plants, animals, and fossils, much of which eventually ended up in the collections of the newly founded Smithsonian Institution.[115] The expedition "can be regarded as a very successful beginning to a program of scientific

109. Henry to Bache, August 9, 1838. (JHP IV p.98).
110. *Congressional Globe.* 25th Cong., 2nd sess., 297. [April 11, 1838]
111. Wilkes Memo, 1838. Reingold, 119.
112. Wilkes to Secretary of Navy, July 16, 1842. Reprinted in Reingold, 124.
113. Goetzmann, 273.
114. Quoted in C. Ian Jackson, "Exploration as Science: Charles Wilkes and the U.S. Exploring Expedition, 1838–1842."*American Scientist* 73:5 (September–October 1985), 455.
115. Philbrick, 332. Originally, the National Institute held the specimens, but after "mismanagement of the expedition collections," they were removed. (JHP VI p.464)

exploration by the United States government."[116] The expedition's claim to fame was its survey of 1,500 miles of the Antarctic coast, one of the first to indicate that Antarctica was a separate continent.[117] "Wilkes Land" can still be found on its maps.

The American voyage had both a positive and negative effect on British efforts to send out their own Antarctic expedition prior to the crusade. During the 1836 effort, Lloyd feared that the admiralty could become reluctant to accede to the proposal of the British Association for a southern expedition, especially as the Americans had taken it up. "The Yankees are preparing an expedition on a grand scale (I understand four vessels of discovery and a frigate)."[118] Indeed, the failure of the earlier British efforts to launch their own Antarctic expedition may have been partly due to the success of the Ex Ex and a desire to avoid redundancy.

However, in the long run Anglo-American rivalry probably helped to propel the British government to approve its own expedition to the southern hemisphere in 1839. American bravado certainly provided a compelling reason for British entry into the profitable Antarctic waters. In 1838, Beaufort wrote to Ross, lamenting that the British government had allowed the Americans to take over 20 million seals from the South Shetland Isles, so much so that the fishery was altogether destroyed.[119] Just as the British government was finally outfitting its own expedition, Bache could already write to Sabine, bragging that "*Our* expedition has been heard from after touching at Rio de Janeiro... They have instruments and mean to use them."[120]

The effects of the Ex Ex were also felt on the continent. Gauss wrote to Sabine that "some days ago I hit in the newspapers on a short notice, that our English navigators have been out-run in approaching to the magnetic South-pole by an American mariner, Captain Wilkes, who is told to have observed in latitude 65" 4'S longitude 147" 30' East."[121] Wilkes had indeed beaten the British to the vicinity of the Antarctic pole. Generously, Wilkes himself even wrote to Ross before the departure of the crusade, advising him on the best course for his upcoming journey.[122] While the crusade left too late to discover Antarctica, it had many other lasting benefits. Yet the British were still galled by their failure to outpace their former colonies. In 1839, Roderick Murchinson of the Royal Geological Society lamented what might have been:

---

116. Jackson, 450.
117. Goetzmann, 281.
118. Lloyd to Sabine, October 31, [18]36. (PRO BJ3/7/86).
119. Beaufort to Ross, May 30, 1838. (PRO BJ2/4/47).
120. Bache to Sabine, April 16, 1839. (PRO BJ3/25/1).
121. Gauss to Sabine, February 3, 1841. (PRO BJ3/2/74).
122. Wilkes to Ross, n. d. [1839]. (PRO BJ2/4/46).

Had the project of an Antarctic expedition been acceded to when it was first proposed...there can be no reasonable doubt, that a discovery of coast, which by its extent may almost be designated that of a Southern Continent, situated in the very region to which its efforts were to have been chiefly directed, must have fallen to its lot; and the flag of England been once more the first to wave over an unknown land.[123]

This comparison between British and American efforts concerning launching a scientific expedition to the southern hemisphere reveals a number of interesting points about the interaction between science and state in the period. British scientists were able to rely upon their aristocratic contacts both within the admiralty and government to push their agenda. The American lobby was forced to make its way through the channels of Congress under a republican system. Unlike the prominence of the members of the British lobby, the American side lacked a central figure around which support could gather. None of the congressional supporters of Ex Ex ever stood out, and his association with the discredited theories of his teacher often hampered its primary scientific sponsor, Reynolds. Indeed, for the American lobby, science was seen as an albatross that had to be downplayed as much as possible. Forced to appeal to a popularly elected body like the House of Representatives, supporters of the Ex Ex had to emphasize the national economic and navigational benefits of the expedition before they could win approval. Even after passage, science remained a problem for the expedition. Although some naval men like Commodore Jones saw science as an integral part of the expedition, they ran up against strong resistance in the bureaucracy that could not be overcome. In the end, Captain Wilkes felt free to dispense with much of the scientific baggage that had been placed on the Ex Ex.

The scientific community received less than they had hoped from the Ex Ex. Gray explained his resignation from the project by stating that he preferred not to accompany an expedition that was "so essentially different from the original" intent.[124] Titian Ramsey Peale, one of the scientists attached to the expedition, later lamented the lost opportunity of what he called one of "the expensive blunders of the past." He charged that in order to save money after the return of the expedition, the scientific corps had been dismissed without even a chance to write up the results of their studies. Peale felt that the items collected by the expedition were only "of value to 'closet-naturalists,' stay-at-home philosophers, and others who could profit by Congressional appropriations of money, liberally made for the care of

---

123. Roderick Murchinson and Edward Sabine, "Address." *BAAS Report* 10 (1840), xxxviii.
124. Quoted in Stanton, 68.

the articles."[125] He also claimed that many of the specimens brought back from the expedition were dispersed, taken in "a general scramble for curiosities...in which some 'Honorable' men thoughtlessly took part."[126]

Herschel's influence is clearly visible when contrasting the British and American efforts. Without a figure like Herschel, the American lobby lacked even a significant patron, who might have been able to rally public support for approval even given the uncertainties of the republican Congress. Another contrast was the form of the resulting expedition, which again emphasizes Herschel's role. The American expedition was more a geographical survey than a scientific voyage; a survey of the South Seas which conducted only temporary observations along the way and collected numerous specimens for later study.[127] The United States lacked the colonial possessions or the imperial desire to set up the system of observatories that formed the second half of the crusade. Additionally, despite the influence of Humboldt in the new world, the United States lacked a figure like Herschel who was ready to use the expedition as a stepping-stone to a greater project of gathering observational data. Indeed, it was not until the 1840s that American scientists began to emulate their British counterparts on this score, at the prompting of some of the same British scientists who had lent their support to the crusade.

---

125. Titian Ramsey Peale, "The South Sea Surveying and Exploring Expedition." *American Historical Record*, 3:30 (June 1874), 249.
126. Ibid., 250.
127. In 1846, Henry himself questioned the need to spend thousands of dollars displaying the expedition's specimens, commenting, "a collection of curiosities at Washington is a very indirect means of increasing or diffusing knowledge." Henry to Hawley, December 28, 1846. (JHP VI p.612).

Chapter Five

An Ample Harvest of Precious Facts

The British expedition launched in the fall of 1839 looked considerably different from those planned in 1835 and 1838. Herschel's role in planning this successful version of the crusade was an important one. The most obvious change was the addition of the four colonial observatories (Canada, St. Helena, the Cape and Tasmania) that had been added largely to fulfill Herschel's plan for systematic collection of data on a global scale. In addition, numerous observatories in Asia (sponsored by the East India Company and endorsed by the lobby) provided even more material. This was a far cry from the "one or two stations" which Lloyd had originally suggested to appeal to Herschel in August 1838, indicating the extent to which science had been able to utilize the resources of the British imperial system for its goals. The observatories of the Crusade were to a large extent a creation of Herschel, and he continued to take an active interest in them for years. Not surprisingly, his involvement in this endeavor did not end with Ross's departure, but continued through the initial stages of data collection to encompass the early results of the project until its first renewals.

In September 1839, the *London & Edinburgh Philosophical Magazine* greeted the launching of the crusade with a full article on "Instructions for the Scientific Expedition to the Antarctic Regions." The review focused primarily on Terrestrial magnetism.

While mentioning the interest in the south magnetic pole (or poles),
the article was more concerned with the wider system of observations
that was to be established to complement the "limited" series already
in progress on the continent. Other secondary topics of interest that
were addressed by the crusade were tides, meteorology, ocean currents
and the figure of the Earth.[1] Thus educated British society was introduced to the crusade, which until then had been the work of only a
handful of devoted scientists. The larger public was also informed of
the undertaking through the *London Times*, leading to one of the more
interesting sequences of events in the history of the crusade.

🕮 CUI BONO?

One of the major differences between the scientific lobbies in the
     United States and Britain lay in their appeal to the public. In the
United States, the lobby had to gain the assent of a (relatively) democratically elected Congress. The Wilkes expedition was openly debated
before Congress long before it reached the executive branch. In Britain,
there was no public debate and no vote before Parliament was ever
taken. While supporters of the Ex Ex had been criticized for planning
the expedition before receiving full congressional approval and funding, the Magnetic Crusade was approved by the sitting government
and was funded by a special grant from the treasury without consulting Parliament. Because of this more aristocratic style of government,
the British project did not have to appeal to public opinion, although
some members of the lobby were interested in doing so. In December
1838, Herschel had written to Whewell to suggest that either he or
Peacock should write a review of the magnetic project for an upcoming
issue of a popular periodical, the *Quarterly Review*.[2] However, it was
not until 1840 that this review (written of course by Herschel) finally
appeared, well after the launch of the crusade. Nevertheless, debate on
the Crusade did receive a public airing in the pages of the *London
Times*. Here the attention drawn by the launch of the expedition also
drew criticism, which eventually challenged not only the success of
the project but also the authority of the scientists who sponsored it.

On September 26, 1839, a letter to the editor of the *Times* appeared,
signed only by a mysterious pseudonym: "Cui Bono." The letter drew
readers' attention to a purported "mistake" in the recently launched
Antarctic voyage. The author began by praising the expedition and
noted the great effort that had been expended to make the voyage of

---

1. "Instructions for the Scientific Expedition to the Antarctic Regions,
   Prepared by the President and Council of the Royal Society." *London &
   Edinburgh Philosophical Magazine*, 95 (September 1839), 177–193.
2. Herschel to Whewell, n. d. [December 17, 1838]. (RS HS 21.272).

the *Terror* and the *Erebus* scientifically productive. He claimed such a voyage of discovery was greater than any "since Captain Cook set out with his old wooden quadrant and beech-bowl compass to explore the Pacific." However, the writer feared that an error had been made which rendered the expedition's principle object of study, terrestrial magnetism, useless.

Cui Bono's concern was with the method used to make the magnetic instruments employed by the crusade for its observations. The supposed problem lay in the poor quality of the "voltaic magnet" used to magnetize the bars in the instruments on the expedition. Cui Bono believed that a magnet made in such a manner was no more than a philosophical toy that was useless for practical observations. Specifically, he charged that instead of having a single permanent north and south pole created at either end, such a magnet had dozens of north and south poles throughout its length. He held that magnets exhibiting such discrepancies could not be depended on as true indicators of polarity since the effect compromised any geomagnetic readings that might be taken. If the magnets of the instruments used by the crusade were of such a type, there would be numerous errors in the observations and calculations of terrestrial magnetism.[3]

The facts of this letter appear to have been based on the statements of one E. M. Clarke, a philosophical instrument maker in the Strand. In a letter dated September 13, 1839 (apparently to Cui Bono and printed later by the *Times*), Clarke had similarly commented upon the poor quality of the magnets used for the crusade's instruments. In the first place he held that they were manufactured from improper steel ("sheer" steel instead of "cast" steel). Secondly, instead of being properly magnetized by steel horseshoe magnets that had been permanently charged from an original lodestone source, Clarke claimed that they were actually magnetized by the voltaic magnets of the Adelaide Gallery of Practical Science. Clarke's conclusion, later echoed by Cui Bono, was that "these apparatus are mere philosophical toys" and that they were totally useless for the purposes for which they were intended.[4]

Although it represented a significant challenge to the possibility of the crusade's success, the attack by Cui Bono attracted little initial attention from the scientific community until Lloyd, on returning from a trip to the continent, took up the issue. He responded with his own memorial to the *Times* dated November 4, 1839. Lloyd defended the magnetization of the bars in London and declared that he had tried the magnets personally when they had arrived back in Dublin. He pronounced them among the best he had ever seen. Lloyd dismissed Cui Bono's claims that the state of the bars precluded good observations,

3. *London Times*, September 28, 1839, 5:3.
4. *London Times*, November 6, 1839, 3:3.

pointing out that the detected strength of the magnetism in the bars disproved the notion that they possessed numerous poles. Lloyd also claimed that a direct examination into the magnetic distribution in bars magnetized in a similar manner to those of the expedition had recently been instituted by some members of a committee of the Royal Society, and that the result had been excellent.[5] Lloyd's answer appeared to satisfy the scientists behind the crusade. Sabine even dismissed the original complaint from Clarke as sour grapes for losing out on the contract for the instruments.[6] However, Cui Bono was neither satisfied nor done with his attack. His next missive not only challenged the instruments but also the authority of the scientific community, and the network of trust that was necessary for the making of scientific knowledge.

On November 15, 1839, a new letter appeared in the *Times*, this time signed more appropriately "Cui Bono?". In it, the author continued his attack on the quality of the magnets used by the expedition. But his special target now was Lloyd himself, who Cui Bono suggested was guilty of trying to cover up his mismanagement of the whole affair. Cui Bono asked why it was that Lloyd had entrusted the magnetizing of the bars to an instrument maker in London, rather than having it done in Ireland where he might have supervised the process himself ("Did he think that the magnetism of London was of a superior quality to what he could procure in Dublin?"). More poignant was Cui Bono's attack on the very reliability of Lloyd's word on the subject. Lloyd had maintained that two out of three magnets from the batch prepared in London for the Antarctic voyage were still being used in his own magnetic observatory, proving their efficiency. But, Cui Bono asked, why should we take Lloyd's word alone for the truth of this statement?

> The days are gone by when *dicti magistri* were received as satisfactory settlements of disputed points. Scientific inquiries in the present age will not be satisfied with loose and general assertions, in which the principle points are left untouched, and no *data* afforded for judgment or comparison.[7]

Here Cui Bono was taking on a central issue of trust in science, that a gentleman scientist could always be taken at his word. Steven Shapin sees this trust as the necessary component of inductive science; the conclusions of a scientist must be trustworthy once it is impossible for each individual to separately verify every single fact.[8] Science had

---

5. Ibid., November 6, 1839, 3:3.
6. "So, after all, the letter in the Times was only the production of a disappointed shopkeeper!" Sabine to Lloyd, November 7, 1839. (RS Te #73).
7. *London Times*, November 15, 1839, 6:4.
8. See Steven Shapin, *A Social History of Truth* (Chicago: University of Chicago Press, 1994).

operated along these aristocratic lines for centuries, but in the age of reform even scientists could be expected to be more democratic. Cui Bono cited Lloyd's claim that a committee of the Royal Society had looked into the matter as an example of this old aristocratic science. He believed that it was his attack which had actually caused the committee to take up the issue in the first place, and held that it was absurd that the judgment of such a group should be flatly accepted just because the members of some committee "hold an inquisition with closed doors into the mysteries of magnetism, and deliver oracular, yet whispered responses to favored ears alone." Cui Bono rankled at the secretive nature of this science. How could the public be expected to believe the word (not to mention fund the experiments) of a few scientists who acted in the dark and furnished no evidence for their claims? "Why, the conduct of the professors of animal magnetism was open dealing, fair play, and wisdom crying aloud from the house-tops, compared to this hermetical style of doing the *experimentum crucis*! Are we returned to the age of alchemists and Royal Rosicrucians?"[9]

Thus the Magnetic Crusade had received its first public debate, two months after the launch of the *Terror* and the *Erebus*. Cui Bono demanded more public accountability from the scientists behind the crusade, especially as they were spending public money on their project.[10] The crusade was not the personal preserve of scientists, but advertised as a national undertaking. As such the public had an interest in whether it was being conducted in a manner that produced valid results for the money spent.

> Practical inquirers of the present day will not be satisfied with statements (which instead of affording the candid explanations they have a right to demand from professional servants of the public) present the appearance of a desire to evade the merits of the case by the aid of hoodwinking, sophistry, and mystery.[11]

"Another violent attack of cui bono," declared Herschel to Sabine the day after the last letter appeared. He left it to Lloyd to decide whether to engage in another round of the debate, which Herschel clearly saw as pointless.[12] Lloyd himself wrote to Sabine shortly after the appearance of the new letter asking for advice whether or not to continue the argument.[13] The overall viewpoint of the scientific establishment was that Cui Bono's attack should not be dignified with a response. Sabine accepted that science should be answerable so far as

9. *London Times*, November 15, 1839, 6:4.
10. In all, the expedition cost £1,324 and operating the observatories £5,039. Sabine to Herschel, April 26, 1841. (RS HS 15.127).
11. *London Times*, November 15, 1839, 6:4.
12. Herschel to Sabine, November 16, 1839. (PRO BJ3/26/85–6).
13. Lloyd to Sabine, November 18, 1839. (PRO BJ3/9/64–5).

public money was concerned, and the public had a right to expect that Lloyd should contradict the assertion that any mistakes had been made. However, having done this, Sabine was of the opinion that "the editor of the *Times* might write his fingers off if he pleased before I would even take the trouble, or condescend, to write a letter for his instruction, or to set right his stupid mistakes."[14]

While Lloyd does seem to have drafted another letter in response (now held by the Royal Society in their terrestrial magnetism archive) he never sent it in to be printed and the debate in the *Times* ended. Lloyd later discovered that Cui Bono was apparently a reporter for the *Times* named Stephens, "by tastes & nature a speculating charlatan!" He learned that Stephens was a friend of Clarke, the philosophical instrument maker who had provided the initial information for Cui Bono's attack, and was convinced that Clarke had engineered the whole affair out of jealousy for not getting the commission to provide instruments for the crusade. In his own final riposte, Lloyd took pleasure in pointing out an amusing conclusion to the whole affair. Earlier he had purchased a "magneto electric machine" from the same Clarke for twelve guineas. Yet now he found that its large compound magnet did not lift a small key! "I presume it is made of [his] *cast* steel."[15]

While it is interesting to see how the issue of the Magnetic Crusade was exposed to public scrutiny, it is also revealing how easily the scientific establishment was able to dismiss Cui Bono's complaints. No supporter of the crusade ever accepted Cui Bono's points as valid, and many saw the whole affair as no more than a cheap publicity stunt (although the initial letter did raise some suspicions among the scientists—Sabine had reported in November 1839 that there was some suspicion that Christie was Cui Bono!).[16] The position of the scientific community was also unassailed, despite the doubts raised by Cui Bono and the possible threat of losing the government's trust. The fact that no public discussion of the crusade occurred until after its departure was indicative of the nature of the lobby in Britain. In the end, Cui Bono's doubts did not affect the results of the crusade, but this correspondence did represent one of the few truly public discussions on the subject.

⚗ LONDON OBSERVATORY

Following the success of the lobby for the crusade, support increased for the erection of an observatory similar to those in the colonies, but in the vicinity of London allowing England to participate

---

14. Sabine to Lloyd, November 21, 1839. (RS Te #76).
15. Lloyd to Sabine, January 8, 1840. (PRO BJ3/10/4).
16. Sabine to Lloyd, November 7, 1839. (RS Te #73).

fully with the continental observations. While Cui Bono may not have had much of an effect upon the crusade, his letters seem to have affected Herschel and his view toward the establishment of this magnetic observatory, providing an example of popular opinion affecting the interaction of science and state. Herschel was especially eager to have an official geophysical station in London. While the Royal Society had made physical observations for years, Herschel felt that the Royal Society was deviating from its true purpose when it meddled "in its collective capacity with the business of practical observation." Laws and theories were the part of science which represented its true sphere; it should not engage in observation. This position reflected Herschel's belief that science consisted of two tiers—a lower level of data collection, and a higher level of theoretical work. The Royal Society clearly belonged to the latter. Herschel believed that a more professional system of observation should be undertaken to replace the current series of irregular observations made by individual members.[17]

By the time of the crusade, a number of magnetic surveys had been carried out in the British Isles, including Lloyd's in Ireland and Sabine's in Scotland. Each of these countries also had its own university-sponsored observatory.[18] There were also a few private English observatories and some magnetic observations were carried out at Greenwich, but as of yet there existed no permanent state-supported geophysical observatory in the country. Without such an emplacement, England remained outside of the German union (*Verein*) of physical observatories into which the crusaders had been so eager to integrate their colonial observations. Herschel confessed that the "want of an English Observatory in full correspondence with the rest has all along been felt to be a defect."[19] Forbes later pointed out the absurdity in establishing magnetic observatories in colonies in both hemispheres but none at home.[20]

In late spring 1840, a proposal for such an observatory in England was making its way through the Royal Society. On June 20 Sabine asked Lord Melbourne for a permanent magnetic and meteorological observatory in the vicinity of London, estimating that it would cost £3,000. Pending the establishment of such a station, Sabine held that a temporary observatory should be established at Woolwich, which he felt was the most eligible site, under the Ordnance.[21] On July 2, John Lubbock reported to Herschel that the Physical Committee of the

---

17. Herschel to Forbes, June 10, 1840. (SAUL msdep7 Incoming Letters 1840 #34).
18. These were Lloyd's in Dublin and Nichol's at Glasgow. Forbes to Nichol, April 25, 1838. (SAUL msdep7 Letterbook II 521–2).
19. Herschel Memo, July 8, 1840. (RS HS 25.6.6).
20. Forbes to Herschel, August 17, 1840. (SAUL msdep7 Letterbook III p.120–2).
21. Sabine to Herschel, June 20, [1840]. (UTX 1093:518).

Royal Society had resolved that a magnetic observatory should be permanently established in the vicinity of London conducting meteorological observations similar to those now made at the Royal Society but on a more extended system.[22]

Herschel was concerned by this sudden proposal for an English observatory in 1840 and expressed his doubts to George Airy, the Astronomer Royal, wishing that the issue could have undergone a wider discussion than it could have received in the Royal Society alone.[23] Herschel claimed that had he known the Royal Society was going to push for an English observatory so soon, he would have said more about it in his article which had just appeared in the *Quarterly Review*. To Lubbock, Herschel wrote that the council had recommended an idea that Herschel did not think was yet mature. "That a resolution of this important nature should have passed off hand—*at a thin meeting of the Council...is*, I think, a pity."[24] From Herschel's point of view, there were still major problems to be worked out before observation could even begin. For example, while the Royal Society had proposed that the observatory be at Greenwich, Herschel felt that it should be on the coast so as to facilitate tidal observations as well.[25]

Herschel held that two or three years' discussion were needed on the topic.[26] He argued that not enough consideration had been given to the planning of the observatory, nor did he believe that the right people were onboard yet.[27] Perhaps affected by the public objections voiced by Cui Bono, Herschel urged caution. Instead of using the informal approach of the crusade lobby, he now argued that an "open and public discussion" was needed to gain approval for an English observatory.[28] He feared that the scientific community was pushing for another government grant too soon after the crusade and was concerned that the proposal had not received enough time or public exposure to gather sufficient support. Herschel also feared that without enough support, the results would be disappointing.[29]

Herschel's theme was that the general subject of a new physical observatory was not ripe and should not have been urged on so soon: "The Royal Society is not wise enough, nor the British Association sober enough for its consideration." Herschel seemed especially concerned that the public would not accept another scientific expense so

22. Lubbock to Herschel, July 2, 1840. (RS MM 11.146).
23. Herschel to Airy, July 6, 1840. (UTX 1054:11).
24. Herschel to Lubbock, July 3, 1840. (RS MM 11.147).
25. Herschel to Airy, July 6, 1840. (UTX 1054:11); Airy to Herschel, August 13–4, 1840. (UTX 1084:627.14).
26. Herschel to Northampton, July 8, 1840. (RS HS 25.6.6).
27. Herschel to Lubbock, June 25, 1840. (UTX 1054:233).
28. Memo, July 8, 1840. (RS HS 25.6.6).
29. Herschel to Forbes, June 10, 1840. (SAUL msdep7 Incoming Letters 1840 #34).

soon after the crusade, or Cui Bono's challenge to it. The public should support the proposal, not have it foisted upon them by scientific authorities. In words reminiscent of Cui Bono, he cautioned that

> the public mind wants preparing for a greatly increased material expenditure in the direction of science and ought to have some guarantee other than the *ipse dixit* of a body (however dignified) or the acclamations of a meeting excited to an extraordinary pitch by the circumstances in which they are brought together that such increased expenditure is desirable on very broad & generally defensible principles.[30]

Herschel wrote nearly identical letters to Sabine, Northampton, Lubbock, Airy and Forbes urging that more consideration be put into the proposed observatory. Unable to attend a meeting of the Physical Committee, he sent a lengthy memo of his ideas. While he confirmed his longstanding support for physical observatories, he was discomforted by the current proposal. However, Herschel was willing to lend some support to the proposal since the application to government had already been made. Herschel seemed to prefer Sabine's plan for an observatory at Woolwich. This would only be a temporary establishment, which could later be replaced by a permanent one of the sort Herschel envisioned after the end of the series of observations carried out by the crusade.

Equally tied up in Herschel's objection was his desire to push another of his long-standing ideas. Just as he had utilized the crusade to aid in the creation of observatories in the British colonies (which he had already proposed in 1838 before the lobby began), now he hoped to use the push for a British observatory as a way to facilitate his notion of creating observatories in conjunction with institutes for conducting experimental physics. In explaining his intent to the Physical Committee of the Royal Society, Herschel noted that the proposal for a new physical observatory should be "connected with an ulterior object of far more importance." That object was a college or institute destined for the systematic determination of the numerical data of experimental science and the coefficients and elements of physical laws.[31] Herschel held that every civilized nation ought to maintain one or more physical observatories for the determination of local data and several experimental institutes for perfecting knowledge that could be improved by experiments systematically performed for the purpose.[32] Both institutions should be state facilities, funded by the government. In this way Herschel addressed the need for state support for both

30. Herschel to Sabine, June 1840. (UTX 1089:632).
31. Herschel to Northampton, July 8, 1840. (RS HS 25.6.6).
32. Herschel to Forbes, June 10, 1840 (SAUL msdep7 Incoming Letters 1840 #34); Herschel to Forbes, August 10, 1840. (SAUL msdep7 Incoming Letters 1840 #42a).

branches of empirical science—observational and experimental. His emphasis on experimental institutes indicates a possible move away from his focus on observational science.[33] Unfortunately for Herschel, his plans were not to be fulfilled.

While Herschel feared that the proposal by the Royal Society might be rushed through without due consideration he also wanted time to sell his idea of an experimental college to be established in conjunction with the observatory. Earlier, Herschel had written to Airy that he had long entertained the idea that a physical observatory and experimental institute or college were desirable as national establishments.[34] As with the Magnetic Crusade, Herschel tried to bring the new proposal for an English observatory into line with plans that had been maturing in his mind for years. Herschel sought to use the Royal Society's proposal for a British observatory to push his own plan for the establishment of national institutions in the form of an observatory and experimental college. He was upset to see the Royal Society pushing ahead on one half of this issue without any attention to the other. Perhaps due to his own interests in establishing an experimental college, Herschel was in no hurry to see an observatory founded. Instead he backed Sabine's idea for a temporary observatory at Woolwich.

While the proposed physical observatory was finally approved, the experimental institute that Herschel had pushed alongside it received no attention. Sabine, though, still saw the English physical observatory as Herschel's idea, writing in 1841 that he had found a very suitable place "for your ulterior project of a Physical Observatory" at the King's Observatory at Kew.[35] An observatory had been established there in 1769 during the reign of George III to observe the transits of Venus. Now under the superintendence of Charles Wheatstone, it became the premier English magnetic and meteorological observatory, as well as the site of the original meridinal line, predating Greenwich.[36] Herschel, however was never happy with his new physical observatory and later disassociated himself from it. Writing to Sabine later in the year, he stated that as a Royal Society establishment, the Kew Observatory "appears to me I confess likely to cause some degree of embarrassment."[37] In 1842 he wrote to Wheatstone, declining to take

33. "The benefits to be expected from such an institution as I have above termed a Physical Observatory, [are] to be quite subordinate in comparison with those pertaining to what I have designated as an experimental college." Herschel to Airy, July 6, 1840. (UTX 1054:11).
34. Herschel to Airy, July 6, 1840. (UTX 1054:11).
35. Sabine to Herschel, February 5, 1841. (RS HS 15.123).
36. The Kew Observatory still stands in the Old Deer Park just south of Kew Gardens in Richmond, Surrey.
37. Herschel to Sabine, December 2, 1841. (PRO BJ3/26/185).

any active part in extending the role of the observatory.[38] Unlike the success of the crusade, Herschel did not get his way with the proposal for the London observatory.[39]

▓ HAMMERFEST

Sabine also found success elusive in his own plan for a new Arctic observatory. Given the prominent role that Hansteen's theories had played in fueling Sabine's interest in geomagnetism, it seemed only fitting that an observatory be founded in Norway to cooperate with the new series of British observations. Sabine broached the subject to Hansteen soon after a magnetic conference in Goettingen in October 1839. Although Hansteen was already constructing an observatory in Christiana (Oslo), Sabine stressed the need for higher latitude observations from the vicinity of the Arctic Circle. As in the British colonial observatories, Norwegian military officers would serve as observers and the British were prepared to pick up the cost of the instruments. Worried that the generosity of the Norwegian government would not stretch to two observatories, Sabine pledged his support in gaining British aid for the project. He suggested Hammerfest near North Cape as the most suitable location.[40]

Since the end of the Napoleonic Wars, Norway had been joined with Sweden under a single monarch. Thus Sabine employed the Royal Society to address King Charles XIV about the possibility of establishing an observatory at Hammerfest.[41] This proposal, however, was misdirected. As Hansteen pointed out, although the Swedish king ruled Norway, it was still an independent nation with its own sense of national pride. "The finances of Norway are quite separate from those of Sweden," he informed Sabine. "The Swedish government cannot dispose of a single farthing in Norway, nor order a Norwegian subject to

---

38. "I have received and read your prospectus relative to the proposed establishment of the British Association at Kew. I see no particular or at least no strong objection to the measure in general, nor to the light in which you have placed it, but I must tell you in candour that I feel no way disposed to take any prominent part in its institution, such as would be implied by my drawing up a report on the subject or altering in any way the draft you have sent me." Herschel to Wheatstone, June 17, 1842. (RS HS 18.149).

39. Similarly, Herschel and William Birt later faced problems gaining state support for meteorological observations which promised few immediate benefits for the expenditure of public funds. Vladimir Jankovich, "Ideological Crests Versus Empirical Troughs." *British Journal for the History of Science*, 31:1 (March 1998), 34.

40. Sabine to Hansteen, October 31, 1839. (ITA).

41. Sabine to Hansteen, December 31, 1839. (ITA).

pick up a pin from the earth"![42] Additionally, the Norwegian governor was abroad, and Hansteen doubted that Charles XIV would act on the observatory without his advice.

As a result of these delays, the Royal Society was surprised to discover that they had been beaten to their goal. In January 1840, the British received word that a French *Commission du Nord* had already established an observatory in the vicinity of Hammerfest.[43] As it would be redundant for the British to place their own Norwegian observatory in the shadow of a French establishment, the Royal Society postponed their plans.[44] However, funding for the French observatory fell through due to a misunderstanding with the French *Institut*, thus leaving the question up in the air.[45] Sabine was eager to combine the French and British efforts, and suggested that the French instruments in Norway be turned over for use in the Hammerfest observatory.[46] Alternatively, he suggested an observatory further south in Trondheim.[47] But in the end the Norwegian government was unwilling to fund another station, and the project had to be abandoned. Sabine regretted this missed opportunity and continued his efforts to establish an Arctic observatory in Norway until at least 1844.[48]

OBSERVATIONAL TIMING

After the successful launch of the crusade, it took some time before all of the fixed colonial observatories in the colonies could be founded. Lieutenant Riddell was responsible for the Canadian observatory. Finding Montreal an unsuitable locality on account of the prevalence of igneous rocks that produced local magnetism he instead chose Toronto as the site of his observatory. This station was completed by September 4, 1840. Lieutenant John Henry Lefroy landed at St. Helena on February 1, 1840, and immediately took possession of Longwood House (where Napoleon had once lived while in exile) as a residence for himself and his detachment. At a short distance from the house, a stone building for an observatory was completed by the end of July. Here, as at Toronto, such observations as could be made in the neighborhood of iron were commenced early in March in one of the rooms of Longwood House, and continued until the transfer of the instruments into the observatory. Lieutenant Wilmot established the

42. Hansteen to Sabine, June 20, 1840. (PRO BJ3/45/65).
43. Sabine to Hansteen, January 21, 1840. (ITA).
44. Sabine to Hansteen, May 29, 1840. (ITA).
45. Herschel to Lubbock, May 31, 1840. (RS HS 22.53).
46. Sabine to Hansteen, October 8, 1840. (ITA).
47. Sabine to Herschel, March 18, 1841. (RS HS 15.126).
48. Sabine to Hansteen, February 3, 1844. (PRO BJ3/30/28).

Cape observatory with the aid of the astronomical observer there, Herschel's friend Thomas Maclear. With the consent of the commanding admiral, a portion of the observatory ground was assigned for the site of the magnetic observatory, and the commanding engineer was directed to proceed with the buildings. Sabine reported that at each of the stations the officers spoke gratefully of the disposition of the governor to give them assistance in the execution of the service on which they were employed. "Their reports are highly satisfactory in respect to the uniform good conduct of the non commissioned officers and men of their detachments."[49]

Ross finally reached Tasmania in August 1840, ten months after leaving England. He and Sir John Franklin, governor of the colony, personally chose the site for the observatory on the banks of the Derwent River. Convict laborers were used to build a stone and timber observatory on sandstone bedrock. The main building was completed and equipped in nine days, a tribute (according to Ross) to "what may be done where the hearts and energies of all are united to promote the common object."[50] The observatory was named "Rossbank" after its founder. Lieutenant Joseph Henry Kay was stationed there as observer, and regularly reported to Ross on the activities of "his" observatory.[51] These colonial outposts regularly sent back material for analysis. The officers manning them fulfilled the position of amateur scientists and data collectors, as envisioned by Herschel's Baconian system.

Prior to sending out the expedition, Lloyd had drawn up instructions for the observers about how and when they should observe. One important issue was that of simultaneous observations. With numerous observatories spread out all over the globe, observations had to be timed so that any variations in the Earth's magnetic field could be compared across the planet's surface. Herschel was of the strong opinion that all of the observations should be simultaneous, as well as linked to those already being conducted on the continent, using the Goettingen meridian as a basis for the time. Again, this point derived from his philosophical system, which required not only observations from across a range of places, but across a range of years. Just as the places of observations had to be known relative to one another, so did the times. The only way that a worldwide collection of observations could be used to determine a general theory was if that data were synchronized. The best results came when all of the data points around

49. Sabine to H.D. Ross, November 10, 1840. (PRO BJ3/27/21-3).
50. Ross might also have attributed the rapid pace of construction to the "energies" of 200 convicts working up to sixteen hours a day. Ann Savours and Anita McConnell, "The History of the Rossbank Observatory, Tasmania." *Annals of Science*, 39:6 (November 1982), 531–532.
51. Kay to Ross, January 1, 1844. (PRO BJ2/712); Kay to Ross, August 10, 1844. (BJ2/7/11); Kay to Ross, November 29, 1844. (BJ2/7/7); Kay to Ross, August 1, 1845. (BJ2/7/2-3).

the world were taken as close together as possible, thus allowing the theorist to compare simultaneous magnetic fluctuations from Toronto to Calcutta. "The only reason of course for this system is the great additional chance it affords (by making almost every observation at some one station strictly simultaneous with *some* observation at *some* other) of pursuing into remote regions and over great geographical amplitudes the traces of simultaneous irregular action."[52]

The issue of observation times was already being discussed before the establishment of the observatories. In July 1839 Lloyd wrote to Sabine that he believed that there should be one strictly simultaneous observation each day, over the whole magnetic field, which would serve to bind together the parts of the system, and to enable them to separate the periodical from the irregular phenomena.[53] However, he could not bring himself to agree with Herschel's proposal that all of the observations should be simultaneous. Lloyd believed that the effect of such an arrangement would weaken the evidence obtained by the regular magnetic changes, and would not yield information of much value regarding the irregular changes. Instead, Lloyd proposed a plan of having one universal field day for magnetism each month and another minor one every week, to be taken in rotation by at least two observatories (one in the northern hemisphere and one in the southern).[54] Sabine proposed a compromise of having one or two simultaneous observations each day over the whole magnetic field, and those always observed by the officers.[55] In the end, though, there were only a limited number of days each year (usually four) when it was practical to carry out simultaneous observations with Europe.

Another point was the interval at which observations were to be made on these days. Here Herschel favored the use of the same five and ten minute "cycle" of observing the three magnetic elements (variation, dip and intensity) that was being used on the continent, again in an effort to tie the British observations as closely as possible to the existing European ones. The original instructions for the colonial observatories and the East India Company observatories designated a two and a half minute interval for individual observations, the multiples of which corresponded to the five and ten minute cycles used by Gauss and Weber, who observed all three elements at once. However, individual practice varied at each observatory, and after several months they made whatever alterations were considered best and thus the intervals came to vary in different observatories from 2½ to 10 minutes.[56] In 1841, Gauss had written to Sabine to point out that in

---

52. Herschel to Lloyd, August 7, 1839. (RS HS 22.25).
53. Lloyd to Sabine, July 2, 1839. (PRO BJ3/9/38).
54. Lloyd to Sabine, July 13, 1839. (PRO BJ3/9/50).
55. Sabine to Herschel, August 1839. (RS HS 15.52).
56. Sabine to Herschel, January 19, 1842. (PRO BJ3/26/191).

general the intensity observations lost most of their value by their being made at intervals much too large (10 minutes) and not simultaneously with the observations of declination.[57] Gauss asked that the cycle on the term days all be changed to five minutes (with two and a half minute intervals). But in August 1841 Lloyd sent supplementary instructions to the observatories changing the intervals to two minutes, making a cycle of six minutes for the upcoming 1842 term.[58]

Herschel's objection to this plan was predictable.[59] A six-minute cycle meant that the colonial observatories were out of correspondence with the European ones that were still on the five-minute cycle. Herschel held that "there ought not *in science* and *in the same nation* to be two reckonings of Time."[60] Writing to Sabine, Herschel pointed out that if they adhered to the alteration in the system they voluntarily put themselves out of simultaneity with all the continental observatories during an operation where strict simultaneity was necessary. Additionally, since Ross had been instructed to use the previous (European) intervals when he set out on his expedition, the change in observation times put the stations out of synchronization with his naval observations as well. Because the European observatories were staying with their existing system of observing, Herschel declared it was unfortunate that the idea of a change was even contemplated. His advice was that the old plan of two and a half minute intervals should be restored.[61]

Lloyd defended the decision to change the timing by pointing out that Gauss had chosen the five-minute interval when there were only two instruments. Now that there were three separate observations to be made, it was difficult to fit them all in. While he admitted that they were likely to forfeit simultaneity with the continental observatories and with the expedition, in his judgment the simultaneity of the individual observatories was of less importance than the shortness of the intervals.[62] Herschel, though, was unconvinced.[63] He was eager to reinstate the old system of observations. In the end, Lloyd assented to Herschel's recommendation and agreed to restore the two and a half

---

57. Gauss to Sabine, May 23, 1341. (PRO BJ3/2/75).
58. Sabine to Herschel, January 19, 1842. (PRO BJ3/26/191).
59. Diane Josefowicz uses the exchange between Herschel and Gauss over the observational timing as the launching point for a discussion of the differences the two men had regarding scientific training and the communicability of scientific knowledge. Diane Josefowicz, "Experience, Pedagogy and the Study of Terrestrial Magnetism." *Perspectives on Science*, 13:4 (Winter 2005), 452–494.
60. Herschel to Physical Committee of Royal Society, February 9, 1842. (RS HS 22.113).
61. Herschel to Sabine, July 12, 1842. (RS HS 15.145).
62. Lloyd to Sabine, July 17, 1842. (PRO BJ3/26/211-2).
63. Herschel to Sabine, August 3, 1842. (PRO BJ3/26/218-9).

minute intervals and the five-minute cycle. In August Herschel wrote to Airy that the return to the original cycle must be considered settled. This decision had to be communicated to each of the stations without waiting for a completion of a code of instructions.[64]

By 1842 Herschel himself admitted that he wished to move on to other projects. "For my own part I can not too often or too strongly declare as I have never ceased to do since this business originated that I am no Magnetician and that my share in its conduct is limited to general advocacy of its objects as a branch of research worthy of national support."[65] Herschel had not seen himself taking on a prolonged role in the crusade. But he had been drawn back into the debate on the timing of the observations because it affected the heart of the system he had helped to establish. Without some simultaneous observations at the stations in correspondence with the European observatories, the British colonial system would be isolated. Just as he had intervened on the behalf of the observatories during the lobby for the crusade, so again Herschel stepped in when Lloyd was contemplating changes that might have disrupted the results of his system.

The observations carried out by the colonial system in the early years after the crusade led to at least one major discovery. Already it had been noted that great magnetic "storms" occasionally swept across Europe, causing dramatic alterations in magnetic variation and intensity.[66] Prior to the establishment of the observatories, Herschel had suggested the possibility that some of the more marked and larger irregular fluctuations should affect the whole surface of the globe. Since these irregular changes of such magnitude were more rare than the lesser ones, it was especially necessary to increase the number of simultaneous observations in order to detect and trace them.[67] The system Herschel set up gave an opportunity to study this event on a global scale.

The first great magnetic storm came in September 1841. Airy later related the events. On the twenty-fifth of that month, a most extraordinary disturbance of the magnetic instruments was observed at the Royal Observatory of Greenwich. The effects were dramatic. Within eight minutes, the declination needle changed its position more than two and a quarter degrees, the vertical force was increased by more than 2.5% and the horizontal force increased about 3.3%.[68] Gauss and the German observers also noticed the event.[69] The disturbances were

---

64. Herschel to Airy, August 25, 1842. (RS HS 22.132).
65. Herschel to Lloyd, August 24, 1842. (RS HS 22.137).
66. The potential extent of the "storms" had been noted in 1825, when a single storm had been detected at both Paris and Kazan in Russia. Whewell, 1857, III:49.
67. Herschel to Lloyd, August 7, 1839. (RS HS 22.25).
68. Airy to Sabine, October 26, 1841. (PRO BJ3/20/10).
69. Gaus to Sabine, March 14, 1842. (PRO BJ3/2/78).

noted in the colonial observatories too. Observations of this disturbance were received from the stations at Toronto and Saint Helena.[70]
Based on their reports, Sabine later stated that a remarkable magnetic
disturbance had occurred on September 25–26, which appeared to have
taken place simultaneously over the whole surface of the globe. The
size and magnitude of the storm was enough to baffle many scientists.
Airy was later to confide to Sabine that "I cannot conceive what was
happening to terrestrial magnetism on September 25 [1841] to make
disturbances everywhere so violent yet so discordant."[71] Only the existence of the global system could have enabled the storm to have been
detected and studied to such a degree.

The observation of the magnetic storm produced the first triumph
for the observing network. Herschel linked the event to his philosophical system, advocating increased observations around the time of
storms, holding that the "storm" observations did tend "to illustrate
these latter changes in a manner so peculiarly salient and striking that
they form what Bacon calls the '*intantive elementes*' or '*luciferce*' of
the Phenomena—instances of that that [sic] sort which spare detail by
conveying the prominent features of the subject direct[ly] to the
intellect."[72] Sabine cited the storm observations as an example of the
work done by the observatories in his year-end report to Major General
H. D. Ross of the artillery,[73] and Herschel saw the success of the storm
observations as ample justification for the whole enterprise.[74] This
achievement gave the crusade hope going into 1842 that it might be
renewed for a further term.

▨ RENEWALS

The initial state sanction for the Magnetic Crusade was for only
three years, at which point the entire project came up for reap-
proval. The first such renewal in 1842 passed with little comment.
Both the continued support from the lobby and the comparably early
stage of the project guaranteed that it would not be shut down prema-
turely. Herschel presented his case for renewal in the 1842 *BAAS
Report*, arguing that continuous observations were vital for the establishment of Gaussian theory and that if they were interrupted at this

---

70. Sabine to H.D. Ross, December 24, 1841. (PRO BJ3/79/39–42).
71. Airy to Sabine, January 10, 1842. (PRO BJ3/20).
72. Herschel to Sabine, February 17, 1842. (PRO BJ3/26/202–3).
73. Sabine to H.D. Ross, December 24, 1841. (PRO BJ3/79/39–42).
74. "If it had produced positively only the single result of demonstrating the
    ubiquity of the 'Magnetic Storms' it would have in my eyes yielded ample
    return for the labour and expense incurred." Herschel to Sabine, January
    23, 1842. (UTX 1054:361).

point, the project would be "arrested in full career."[75] He had already
prepared the ground for his argument the previous year in a letter to
Sabine, calling the geophysical project the best scientific use of state
resources that he could imagine. Because the observations had to be
carried out over a long period of time, renewal was essential.[76]

Lobbying for renewal began in earnest in January 1842. Herschel
wrote to Sabine, pointing out that they really did not have three years
of observations from any of the stations, as so much time had been
consumed in preparation. Additionally, a great many new continental
observatories were just coming into activity (especially in Russia).
Herschel also made reference to the recent confusion over the timing.
He saw the first period just elapsing as experimental and preparatory.
With a new period, a thorough and complete revival of the system of
observation could be made and a perfect understanding come to with
the continental authorities (Gauss, Weber, etc.) as to the times of observation.[77] Not all members of the former lobby were convinced of
the need for a new term of observations. Airy, the astronomer royal
was a prominent negative voice.[78] Peacock agreed with Airy in thinking that the committee was not justified in urging upon the government the resumption or extension of the magnetic observatories,
unless there was very important and definite work to be done by them
such as could not be done by other means.[79] Overall, though, most scientists supported the renewal.

In May, Northampton was again employed to make the case to the
government for the continuation of observations. The renewal was
granted with little of the discussion or drama that had marked the
original lobby. Already by June 1842 Lloyd heard that the government had sanctioned an additional term.[80] Gauss had heard the news
by fall.

---

75. "Report of the Committee." *BAAS Report* 12 (1842), 3–4.
76. Herschel to Sabine, April 27, 1841. (RS HS 15.128).
77. Herschel to Sabine, January 23, 1842. (UTX 1054:361).
78. Airy's opposition may have been due to personal disappointment. In May
1842, Airy had written to Northampton asking him to apply to the government on Airy's behalf for an increase in funding (from £550 to £670) for
the magnetic observations being carried out at Greenwich. In June, he had
written to Sabine "I was very much mortified to find that Lord
Northampton had not made application for the continuation of the
Greenwich Magnetic Establishment. It ought most certainly to have gone
with the other application. In regard to support, it is on precisely the same
footing as the Ordnance observatorie: both are supported by special grant
of the Treasury...But my letter was mislaid!! I can endure this once, possibly twice." Airy to Northampton, May 16, 1842. (PRO BJ3.20.30–1); Airy
to Sabine, June 6, 1842. (PRO BJ3/20/32).
79. Peacock to Herschel, May 22, 1842. (UTX 1060:1206).
80. Lloyd to Sabine, June 9, 1842. (PRO BJ3/11/77).

> I need not tell you with how much interest and pleasure I have learned
> from your favour, that the magnetic observations established abroad by
> the munificence of the British Government are to be continued for a
> fresh period of three years. I have no doubt that by this means an ample
> harvest of precious facts will be collected, whence science shall derive
> new progress.[81]

Of more interest was the renewal of 1845. By this point, the results
of the project were beginning to be known and an overall evaluation of
the initial success of the Magnetic Crusade was possible. Herschel
himself had been unsure in 1844 whether the observations should continue for another term. He did not regard the "mere accumulation of a
mass of observations, however accurate and continuous," as constitut-
ing the scientific result that he had in mind. As per his philosophical
system, data collection served only to supply the "real" work of science: the development of theory. Herschel believed that a pause to
allow for reduction of the existing observations might be in order, noting that a majority of the observations which had been received were
not yet in Lloyd's hands and hence had not been subjected to any theoretical examination whatever.[82] Indeed, Herschel was wary of merely
"graphing data for data's sake."[83] He concluded that under these circumstances, an interval might well be devoted to the discussion of results and the direction the project should take.

Sabine, however, was able to convince Herschel to support another
term of three years for the observatories. He pointed out that the
Tsarist government had made arrangements for carrying out the proposed system of research through the whole extent of the Russian empire, and that while the Russians did not afford the same facilities as
England did, they were ready to maintain the system in concurrence
with England until their objects should be accomplished.[84] In addition,
Sabine had the ear of the new prime minister, Sir Robert Peel, who had
remarked that three years seemed a very short time for the accom-
plishments of the objects for which so extensive a system had been
called into operation. Indeed, Peel seemed to have no objection to ap-
proving another three-year period of observations, if that was what the
scientists asked.[85] Eventually, Herschel agreed to address the government to continue the existing observatories in activity until 1848,
with that to be regarded as the final term. Only high expectations of

---

81. Gauss to Herschel, October 24, 1842. (RS HS 8.78).
82. Herschel to Sabine, July 22, 1844. (PRO BJ3/26/254–5).
83. Quoted in Thomas Hankins, "A 'Large and Graceful Sinuosity': John
    Herschel's Graphical Method." *Isis* 97:4 (December 2006), 625.
84. By 1840, the Russians had promised to establish nine observatories to act
    in concert with the British series of observations. Sabine to Hansteen,
    May 29, 1840. (ITA).
85. Sabine to Herschel, July 19, 1844. (PRO BJ3/26/252–3).

considerable scientific results for the future might justify a further extension.[86]

For this renewal, Herschel decided to poll the scientific community to discover their opinions about the results of the project. He asked for written responses from a number of individuals on whether there were any important objects to be accomplished by a continuance of the existing establishments for a longer period and whether private research had been stimulated by the example of the government establishments in Europe and elsewhere.[87] This poll brought opinions from a number of the scientists who had been participating in the various branches of the project. While they expressed varying views on the overall success of the undertaking, the nature of the poll allowed Sabine to finesse some of the answers that Herschel received to ensure that the project had the best chance of renewal. For example, when Quetelet's response failed to reach his requirements, Sabine suggested that Quetelet look over the recently published observations from Toronto, and update his opinion with a postscript. Sabine indicated that several others in the poll had waited until they had examined the Toronto volume, which he felt presented a fair account of the manner in which the proposed objects at the British colonial observatories were carried out.[88]

There was also a Magnetic Congress planned for the meeting of the British Association in 1845 to which Gauss, Humboldt, Kreil, Plana, Hansteen, Quetelet, Weber and Bache had been invited.[89] This conference recommended that the British colonial observatories be continued until 1848 and that the East India Company stations be renewed as well. Optimistically, the attendees hoped the cordial cooperation that had hitherto prevailed between the British and foreign observatories, having produced important results and being absolutely essential to the success of the great system of combined operation that had been undertaken, continued to prevail.[90]

Declaring, "terrestrial physics...worthy to be associated with Astronomy," Herschel now left no doubt as to his opinion.[91] Supporting the renewal, he made his case in the 1845 report of the British Association. Herschel claimed that the Magnetic Crusade had resulted in the greatest system of combined observation ever witnessed in the world. Such an amount of data could not help but advance science. Because of the nature of the work and the promise of more to come, it could not be interrupted.[92] The project had enjoyed a generous level of

---

86. Herschel to Sabine, July 22, 1844. (PRO BJ3/26/254–5).
87. Circular Memo, November 1844. (UTX 1089:645).
88. Sabine to Quetelet, March 8, 1845. (APS HS #11).
89. Herschel to Whewell, July 22, 1844. (RS HS 22.201).
90. "Resolutions," June 25, 1845. (PRO BJ3/26/345 & 352).
91. "Report of the Committee." *BAAS Report* 15 (1845), xxxiii.
92. Ibid., xxx.

government support that should also continue. After the objects of national defense, the due administration of justice, and the healthy maintenance of the social state, Herschel held that there was "no object greater and more noble—none more worthy of national effort—than the furtherance of science."[93] Again, Herschel tied science to civilization. No nation calling itself civilized could deem its institutions complete without the establishment of a permanent physical observatory.[94]

Herschel was joined by scientists from around the world in support of Britain's continued support of the magnetic project. In response to his poll, many wrote with their support for the project. Wilhelm Weber felt that it was impossible to think of interrupting the observations at this point, and cited the indirect results which the systematic prosecution of magnetic observations may have had in exciting and furthering other scientific efforts. He held that magnetic observations were therefore not only necessary for terrestrial magnetism, but had also become an important element for many physical observations.[95] Elias Loomis of New York University agreed, calling the abandonment of observations at this point prejudicial to science. To stop now would be to stop short in the race when the prize was just within reach.[96] "I cannot but look upon accumulating treasures of the Magnetic crusade with the deepest interest, and trust the observations will not be permitted to stop until the victory is fairly won."[97] Johann Lamont in Munich hailed the general cooperation between nations that the project had engendered, which he believed had not at any former period been so conspicuously manifested.[98]

However, not all voices were favorable to renewal. J. A. Brown, the assistant observer at the Makerstoun station, held that regular term observations were (at least at present) unnecessary. While there was little doubt that magnetic disturbances occur for the most part simultaneously over the whole world, what was the use of prearranged periods of observations, when the observer should and should not observe?[99] John Phillips saw no effect of stimulating private experimental research from the observatories.[100] George Airy agreed, citing the observatories' failure to generate private research as a reason to discontinue regular observations at the end of the term. Echoing Herschel's earlier reservations, Airy felt that there was a greater need to reduce the existing observations than to continue to collect more.

---

93. Ibid., xxxii.
94. Ibid., xxxiv.
95. Ibid., 17.
96. Ibid., 20.
97. Loomis to Sabine, March 31, 1845. (PRO BJ3/25/56-7).
98. "Report of the Committee." *BAAS Report* 15 (1845), 23.
99. Ibid., 34.
100. Ibid., 37.

They needed time to calmly compare the results. As long as the observations should be continued "so long *at least* will the real intellectual progress of the science be put off."[101]

Despite some reservations from within the scientific community, the magnetic project was renewed. Herschel once again turned to Northampton to ensure timely action. When no reply had been received from Robert Peel on the subject of the magnetic observatories by October, Herschel wrote asking Northampton to intervene. Time was pressing for news of the renewal to go out to the stations. If nothing definite was ordered by the end of October 1845, even though the reply should be ultimately favorable, it was possible that the continuity of the observations would be broken.[102] In the end, the lobby was successful and news of the renewal came by the beginning of November.[103]

▩ HERSCHEL'S DEPARTURE

The 1845 renewal marked the end of Herschel's direct involvement with the project. He must have been reassured by both the government renewal as well as the East India Company's decision to make its Indian observatories permanent in 1845.[104] Following the efforts he undertook for the 1845 renewal, Herschel retired from the magnetic project. Writing to Sabine earlier, he explained his decision:

> I cannot under take in future to devote to this subject that share of my time and attention which I have hitherto found it to require...My scientific life draws towards its close. I have much work upon my hands, more than I can hope it will be possible for me to execute, and duties of a widely different nature and whose demands grow annually more urgent, abstract already many hours daily of the time which I could at an earlier period devote to studies & whether consonant to my own pursuits or not.[105]

By 1845, Herschel felt that the subject of geomagnetism had "fairly outgrown the moderate amount of knowledge of it I ever possessed," and that he, Peacock and Whewell had already been reduced to the status of "sleeping partners" in the project anyway.[106]

Herschel could take pride in the extent to which his initial goals had been achieved. His plan to use British colonies as observing bases for inductive science was now a reality. In addition to the four stations set

---

101. Ibid., 52–53.
102. Herschel to Northampton, October 8, 1845. (RS HS 22.250).
103. Trevelyan to [...], November 1, 1845. (PRO BJ3/27/246).
104. EIC to Herschel, November 4, 1845. (RS HS 17.142).
105. Herschel to Sabine, July 22, 1844. (PRO BJ3/26/254–5).
106. Herschel to Lloyd, May 30, 1845. (RS HS 22.236).

up by the crusade, other colonies were joining the network. British governors in Ceylon, New Brunswick, Bermuda and Newfoundland had asked that physical observatories be established in their respective colonies.[107] In 1845 British Guiana set an example which Herschel hoped "all our colonies will follow," by establishing (at their own expense) a permanent physical observatory and setting aside £100 a year to employ an observer.[108] Sabine felt that the establishment of colonial observatories should have the effect of stimulating similar institutions in the wealthier colonies, working to the direction of the overall project.[109] For Sabine, the "importance of encouraging a disposition in the Colonies to participate in the interest excited by great subjects of scientific enquiry, and to contribute their portion to the advancement of knowledge, [wa]s too obvious to be dwelt on."[110]

British science and its connection to the state had taken on a new form in the period 1830–1845. Scientists now found themselves in a position where they could expect government aid for their projects. The success of the crusade helped to establish a new relationship between science and state, tying together scientific and imperial interests. The continuity and benefit of state sanction contributed greatly to projects such as the Magnetic Crusade. Fields of science which could not have been studied without state support now opened to exploration. Herschel's role in this transformation was both significant and essential to the shape of British science in the nineteenth century.

---

107. Sabine to Herschel, April 11, 1845. (PRO BJ3/30/54).
108. Herschel to Sabine, May 1845. (PRO BJ3/26/335–6).
109. Sabine to Herschel, June 2, 1845. (UTX 1093:521.17).
110. Sabine to Herschel, June 17, 1844. (PRO BJ3/27).

Chapter Six

Knowledge and Philanthropy Among the Nations of the Earth

Just as the crusade continued to inspire research and activity in Britain, so its impact was felt across the Atlantic. The establishment of an observatory in British Canada encouraged new attempts to involve the state in scientific activity in the United States (although not without continued political wrangling). Magnetic and geological surveys opened the way westward for American imperial expansion across the continent, while Herschel's system of colonial observatories became the pattern for American meteorological observations. The Toronto observatory became the center of a ripple effect, inspiring the foundation of American observatories to participate in the British colonial series of observations. As at the Cape and other locations, these outposts could themselves become important centers of scientific research. Eventually, American financial support became an important part of the continued operation of the Toronto observatory. In the end, the imperial setting allowed continuous observations on a scale never before attempted. These observations were crucial to Sabine's eventual discovery of a "cosmic" link between solar and terrestrial magnetism, helping to explain the aurora about which Henry and Loomis had wondered years before. It was fitting that the final results of the Magnetic Crusade were to come back to the universal view of the world that had first inspired it, demonstrating that phenomena such as terrestrial magnetism could not be fully studied on a local, or even a planetary, level. The extension of scientific study across the empire was only the beginning.

✻ AMERICAN REACTIONS TO THE CRUSADE

As in Britain, the geophysical sciences continued to be a major subject of study in the United States. American geomagnetic and meteorological research flourished after the launching of the Ex Ex, influenced especially by British steps in the same direction. Eventually the two sides made efforts to pool their resources, and American science was brought into the global system created in the wake of the crusade. The American press and scientific community warmly received the advent of the Magnetic Crusade. The official announcement from the Royal Society reached the United States by July 1839.[1] But even in April, while Herschel was still concerned that the expedition would be held up because of the lack of equipment for the stations, Alexander Bache had already heard of the "magnetic news" in Philadelphia.[2] The American press announced the crusade from Georgia to Maine in June.[3]

With the launching of the expedition there was a concerted effort on the British side to involve as many countries as possible in the second half of the crusade, the establishment of observing stations. The British scientists hoped that in addition to the stations established in their current colonies, their former American colonists would also take part. In September 1839, Lieutenant Riddell, assigned to the new Canadian observatory, met with Alexander Bache in Philadelphia and Joseph Henry at Princeton. Bache was greatly impressed by the extent of the British plan, and Riddell reported to Sabine that the Philadelphia observatory would endeavor to conduct hourly observations (or at least to take part in the monthly term days as often as possible).[4] Bache later wrote to Renwick concerning Sabine's suggestion that the United States set up its own magnetic survey.[5] British scientists were eager to integrate American stations into their system and occasionally intervened to try to aid their American counterparts. Herschel wrote to Paine that:

> American stations are of the utmost importance and we will hope that an appeal to your great and energetic country will not be lost but that America will supply her quota to this grand accumulation of data by

---

1. Smyth to Henry, July 1, 1839. (JHP IV p.243).
2. Bache to Sabine, April 16, 1839. (JHP IV p.203).
3. Henry to Ross, June 19, 1839. (JHP IV p.236).
4. Riddel to Sabine, September 23, 1839. (PRO BJ3/34/1); Bache to Sabine, June 29, 1840. (PRO BJ3/25/7).
5. Bache to Renwick, September 30, 1839. (JHP IV p.265).

which the actual magnetic state of the globe will be fixed for future ages so to speak at a blow. The opportunity lost can never be recovered. So many favorable conditions can never be again expected to conspire.[6]

Lloyd also supported American observations, and urged the Royal Society to petition the American Philosophical Society to find funds to supply Bache with an assistant for observing.[7]

In December 1839, Henry felt confident that the American government could be convinced to establish a series of magnetic and meteorological observatories on a plan similar to that of the British colonies.[8] A committee of the American Philosophical Society (including Henry) petitioned Secretary of War Joel Poinsett to establish such observatories, inviting his department's cooperation in the extensive system of magnetic and meteorological observations about to be made under the direction of the British government. The society favored the establishment of five magnetic observatories: one each in the northeastern, northwestern, southeastern, southwestern and central part of the United States.[9] Herschel embraced the possibility that the American proposal offered to extend his system, recommending the full support of the Royal Society: "There can be no sort of question that the grand combination in hand will be deficient in one of its most important features if American observations be wanting or rather if they be not abundantly supplied."[10]

Unfortunately, the American government proved less responsive than the British to the idea of establishing observatories. The American Philosophical Society's proposal, brought before the Democratically controlled House of Representatives by former president John Quincy Adams (Whig, MA) was defeated overwhelmingly by a vote of 34–97 in July 1840. The Whigs split almost evenly while nearly all of the Democrats opposed the observatories.[11] Elias Loomis was gloomy about the outcome, and feared that governmental support for science would not be forthcoming, leaving scientists to their own designs. "Our own Government seems to feel very little interest in the cause of science," he complained. "The election of a President is a matter of far greater importance than the discovery of the laws of magnetism."[12] A

---

6. Herschel to Paine, November 4, 1839. (UTX 1054:292).
7. Lloyd to Sabine, March 20, 1840. (PRO BJ3/10/35).
8. Henry to Henslow, December 2, 1839. (JHP IV p.310).
9. American Philosophical Society to Poinsett, December 20, 1839. (JHP IV p.315).
10. Herschel to Sabine, April 29, 1840. (PRO BJ3/26/116–7).
11. In favor were thirty-two Whigs and two Democrats. Twenty-five Whigs and seventy-two Democrats were opposed. U. S. *House Journal*. 1840. 26th Cong., 1st sess., 18 July 18.
12. Loomis to Sabine, October 3, 1840. (PRO BJ3/25/9). Loomis was referring to the presidential campaign of 1840 between Martin van Buren and William Henry Harrison. The result was a Whig victory.

number of private observations were carried out, including those at Bache's observatory at Girard College in Philadelphia. His efforts corresponded with those established by the British for the crusade but were supported through a private subscription by members of the American Philosophical Society.[13]

In its failure to convince the government to go along with the British plans for observatories, American science ran into the barrier that hampered the relationship between science and state in the United States throughout this period. More so than in Britain, the cause of science was hard to sell in the young republic. *Laissez-faire* political beliefs and anti-international traditions stood in the way of the successful cooperation of science and state that was demonstrated by the Magnetic Crusade. That is not to say that there was no such cooperation, only that it took on its own American form. While the federal government might not be willing to embrace a large-scale scientific project, some of this deficit was made up by the state governments, which were often more inclined to support science. American science still had to work to present itself as professional and worthy of federal support.

The states were particularly interested in geological surveys to find mineral deposits and railroad routes. Many states followed the example of North Carolina's geological survey of 1823, the first in the nation. In the 1830s, fifteen more surveys were funded.[14] When the government of New York ordered a geological survey of the state to be carried out in 1836, Henry pushed the state to include geomagnetic observations as part of the survey, but was unable to convince the legislature to go along with his plan.[15] However, he did receive assurances that such a survey would be carried out either by the board of regents of the university or as part of the geological survey if there were any spare funds.[16]

In 1841 Henry summarized the state of scientific research in the United States. He complained that although the republic had more persons interested in popular science "than in any other part of the world," few were engaged in original research. Geology had become a "very fashionable" science, inspired by the vast unexplored lands to the west. Two geomagnetic observatories were then in operation, one in Boston and the other in Philadelphia. The nation also supported two series of meteorological observations, one in the academies of the state of New York (which Henry had a role in establishing) and the other

---

13. Henry to Petty Vaughan, March 30, 1840. (JHP IV p.342); Bache to Sabine, July 21, 1841. (PRO BJ3/25/13).

14. Robert Bruce, *The Launching of Modern American Science*. (New York: Knopf, 1987), 166–170.

15. Henry to Forbes, June 7, 1836. (JHP III p.73).

16. Henry to Bache, May 10, 1836. (JHP III p.60).

carried out at various military posts by order of the army.[17] While the immediate effect of the Magnetic Crusade was to encourage American scientists to take part in the global research they saw being carried out through the British domains, it would be a while before American contributions matched the British effort.

CANADIAN SURVEY

The establishment of a colonial observatory in Canada brought British geophysical interests closer to the United States. Originally intended for Montreal, the observatory was eventually located in Toronto because of magnetic interference at the former site. Lieutenant Riddell was the first artillery officer appointed to supervise the observations. He suggested the idea of conducting a magnetic survey of nearby regions once the observatory was up and running.[18] Sabine then became interested in the idea of a full magnetic survey of the whole of the British possessions in North America, using the Toronto observatory as a primary station for reference and comparison.[19] In 1841, he proposed a magnetic survey of Canada, estimating a cost of £1,170 for the whole project.[20] His proposal was accepted and the survey was ordered. Already by August 1841, Loomis greeted the news of the British magnetic survey of Canada with anticipation, calling it "the most important point in the Northern hemisphere."[21] As planning for the survey continued, Riddell returned to England and was replaced at Toronto by Lieutenant Lefroy, who had earlier manned the St. Helena observatory.[22]

Surveying the vast regions of Canada required the cooperation of that other great British trading enterprise, the Hudson's Bay Company, whose zone of control still stretched across most of western Canada.[23] In 1843, the Royal Society requested passage for Lieutenant Lefroy through the land occupied by the company.[24] Fortunately, the Hudson's Bay Company proved to be as willing to aid the cause of science as their East Indian counterparts. George Simpson provided the necessary letters of introduction for Lefroy, "to proceed to any part of the country he may desire, and to make such stay at the different posts as he

---

17. Henry to de la Rive, November 24, 1841. (JHP V p.119).
18. Herschel to Whewell, April 12, 1840. (RS HS 22.48).
19. Sabine to Herschel, November 25, 1840. (PRO BJ3/26/157).
20. Sabine Memo, April 15, 1841. (PRO BJ3/27/42).
21. Loomis to Sabine, August 10, 1841. (PRO BJ3/25/17–8).
22. Lieutenant Smythe replaced Lefroy at St. Helena.
23. The Hudson's Bay Company ceded control of most of its Canadian land in 1870.
24. Barclay to Sabine, December 21, 1844. (PRO BJ3/27/195).

may."[25] The original intent was for Lefroy to accompany a heavy brigade from Montreal to Norway house. Lefroy requested two extra observers to aid him in surveying the area through which they would travel. Lefroy set off on May 1, 1843. However, the brigade soon found that Lefroy was slowing their progress too much. The officer in charge complained that the observations took so much time that the brigade would be unable to reach Norway house until too late in the season. Several days later Simpson, traveling separately by canoe, overtook the brigade in time to settle the argument. He suggested that Lefroy take four additional men and set off separately by canoe with guides provided by the company. In this way, Lefroy was able to take whatever time he needed for his observations, and could venture independently to places that the brigade might not cross in order to make additional observations. Lefroy accepted this solution.[26]

Lefroy proceeded to Norway house by his own route. From there he went on to York Factory and then returned to Athabases by way of Norway house. While Lefroy carried out this initial leg of the survey, Sabine was in correspondence with an American scientist, John Locke of Ohio, suggesting that he conduct a series of observations which would complement Lefroy's. "It is much to be desired that this Survey should be met at the Frontier by researches of Citizens of the United States conducted on their own grounds."[27] Sabine wrote to Lefroy over the winter of 1843–1844 to inform him of the observations that Locke was conducting along the states of the northern frontier. Sabine was excited to note that Lefroy's observations seemed to support his own theory that magnetic intensity decreased as one moved north. He even believed that with sufficiently delicate instruments precise latitude could be determined through this phenomenon.[28]

In the spring of 1844 Lefroy set out again on a fourteen-month journey that took him to the Great and Lesser Slave Lakes, Mackenzie and Peace Rivers and Saskatchewan.[29] This plan significantly augmented the original survey planned by Sabine, and incurred significantly higher costs, leading the Hudson's Bay Company to point out that "it was not intended, nor could it be expected, that canoes and men should be withdrawn from the services and appropriated to his [Lefroy's] exclusive use, at the Company's expence."[30] However, the differences were later smoothed over. In 1847, an official at the Hudson's Bay Company commented to Sabine that they were

---

25. Simpson to [...], April 24, 1843. (PRO BJ3/27/201–2).
26. Simpson to Lefroy, May 12, 1843. (PRO BJ3/27).
27. Sabine to Locke, November 22, 1843. (PRO BJ3/30/25).
28. Sabine to Lefroy, March 29, 1844. (PRO BJ3/40/32).
29. Simpson to [...], January 14, 1845. (PRO BJ3/27/197–8).
30. Barclay to Sabine, December 21, 1844. (PRO BJ3/27/195).

always happy to have it in their power to promote scientific objects [and were] much gratified to find that the facilities afforded by the Company to Captain Lefroy have enabled you to make important contributions to terrestrial magnetism.[31]

In addition to tensions between Lefroy and the company, the lieutenant was also aggravating his superior officers, insisting that his scientific responsibilities should release him from other military duties. In 1845 Lefroy apparently had an altercation with one Lieutenant Colonel Maclachlan when Lefroy neglected an order from Maclachlan. This led to a censure and a demand for an apology from Lefroy for his "highly unbecoming and improper" conduct lest he be recalled from service. The division between duty to science and duty to state quickly became apparent. The master general declared that "temporary employment of either officers or soldiers of the Royal Artillery on the magnetic Service does not convey to them either altogether, or individually any such *'independence'* of their superior officers as Lieut[enant] Lefroy has attempted to assert."[32] Lefroy apologized. Despite the tensions created between his scientific agent, the Company and the Artillery, Sabine was pleased by the success of the Canadian survey. He pointed out the benefits of Lefroy's survey for the British navy's upcoming expedition to search for the Northwest Passage, and suggested that Lefroy offer advice on magnetic observations that could be made on this voyage.[33] In 1845, Lefroy suggested a continuation of the Canadian survey to Sabine, estimating a cost of at least an additional £1,420.[34]

The resolutions passed by the British Association at the time of the 1845 renewal included two items of interest for Canada. The first asked that the Canadian survey be continued until observations at the Toronto observatory were connected with a (as yet nonexistent) system of American geophysical observations. Sabine admitted that "this is purposely indefinite; and will justify the continuance of the Survey for 3 years more." A second resolution proposed a new survey to the northeast or northwest of the Hudson's Bay Company's holdings.[35] By 1847, Sabine hoped to extend the survey from the northern polar sea to the Rocky Mountains in the west, and to join the Canadian survey with observations by Franklin's expedition in the arctic and American observations further south to the Gulf of Mexico. The stage was set for a massive, continental system of observations involving British and American science.[36]

---

31. Barclay to Sabine, May 13, 1847. (PRO BJ3/27/319).
32. H. D. Ross to Sabine, April 12, 1845. (PRO BJ3/27/228).
33. Sabine to Lefroy, December 24, 1844. (PRO BJ3/40/58).
34. Sabine to Lefroy, February 28, 1845. (PRO BJ3/40/63).
35. Sabine to Lefroy, July 1, 1845. (PRO BJ3/40/70–1).
36. Sabine to Northampton, May 18, 1847. (PRO BJ3/30/72–4).

▨ AMERICAN GEOPHYSICAL DEVELOPMENTS

As the Canadian survey was continuing, the American geophysical sciences were undergoing their own development, albeit more gradually and with less state support. In 1846, Henry complained that the United States still gave less encouragement to original research than any other civilized country. He compared the few rewards offered in the United States to his idea of the British system, whereby a distinguished scientist could look forward to a university fellowship, church living or even a government posting. Henry felt that while Americans continued to put too much emphasis on practical applications, "the discovery of a new truth is much more difficult and important than any one of its applications taken singly."[37] Now in charge of the Smithsonian Institution, Henry was committed to its goals of encouraging original research and the increase of human knowledge. Reviewing the projects that he lent Smithsonian backing, he praised the original research in the manner of the British Association, especially concerning magnetism and meteorology.[38]

Geographical surveys continued to give American science opportunities for new investigations. James Graham, surveying the boundary line between Vermont and Canada, also took magnetic readings along the way.[39] In 1847, Henry was able to convince the American government to add magnetic observations under the direction of the Smithsonian to its planned surveys in the new territory of Wisconsin.[40] The end of the Mexican War created a new boundary to survey, as well as opening a vast region of relatively unexplored land to American science.[41] Leaving the Canadian frontier, James Graham looked forward to joining the army in Mexico, hoping to contribute something to magnetism and meteorology "from a quarter which has as yet been but very little explored."[42]

There was a general agreement among many scientists that the Mexican lands offered a magnificent opportunity for science. Others saw science as the good that would come out of the evil of war. Francis Lieber wrote to Henry, asking that the American government find some scientific inquiries to pursue in Mexico, to "set a stamp of civilization, or an additional one, upon the rude work of war." He hoped that American soldiers in Mexico could contribute as much to science as

---

37. Henry to Bache, September 5, 1846. (JHP VI p.493).
38. Henry to Hawley, December 28, 1846. (JHP VI p.610).
39. Graham to Sabine, December 3, 1845. (PRO BJ3/25/72).
40. Henry to Loomis, April 22, 1847. (JHP VII p.84); Henry to Sabine, August 13, 1847 (PRO BJ3/32/6–7).
41. See Bruce, 204–205.
42. Graham to Sabine, March 29, 1847. (PRO BJ3/25/128–9).

Napoleon's had done in Egypt.[43] Samuel Haldeman felt that the boundary survey provided an excellent chance of advancing terrestrial magnetism, the "region to be traversed being unknown in the departments of nature science."[44] Henry himself arranged for magnetic instruments to be shipped from Britain for use on the boundary survey.[45]

But it was into the field of meteorology that American science poured most of its geophysical interest in the mid-nineteenth century. Knowledge of even basic meteorological data was scarce in the republic. As late as 1849, Henry had to admit that he did not even know the average annual rainfall in the United States.[46] Independent records of meteorological observations had been kept at various places throughout the country in preceding years. In the 1840s, a new push to create a national system of observations was underway. In his centennial address to the American Philosophical Society in 1846, Loomis charged the society with setting up a system of hundreds of meteorological observers, calling for "a general meteorological crusade" in answer to the British Magnetic Crusade. Their efforts could bring about the day when "men would cease to ridicule the idea of our being able to predict an approaching storm."[47]

At the center of this effort was James Espy, known as the "storm god" to his students, who had spent many years working on a theory of storms.[48] Espy's theory of how storms were naturally created led him to believe that artificial storms were also possible.[49] He held that hot, moist air rose and condensed to form clouds and rain. With the right conditions, one could induce rain by creating an updraft of hot air (for example, by controlled forest fires).[50] He proposed a rainmaking plan whereby thousands of acres of timber would be set afire in the western United States during each summer, which he believed would result in a line of storms forming along the fires and moving east. This exercise in weather control would not only provide beneficial rains to the eastern half of the country, predicted Espy, but would eliminate the problems associated with cold weather, heat waves, famines and epidemic diseases originating from floods. Through his scheme, "the health and happiness of the citizens will be much promoted."[51]

---

43. Lieber to Henry, October 12, 1847. (JHP VII p.199).
44. Haldeman to Henry, March 13, 1849. (JHP VII p.490).
45. Henry to Sabine, November 7, 1849. (JHP VII p.632).
46. Henry to Ellet, May 28, 1849. (JHP VII p.539).
47. Elias Loomis, "On Two Storms Which Were Observed..." *Transactions of the American Philosophical Society*, 9 (1846), 183–184.
48. Letter, June 23, 1845. (DU Patterson).
49. Letter, March 6, 1842. (DU Evans).
50. James Espy, *The Philosophy of Storms* (Boston: Little and Brown, 1841), 492–518.
51. Espy Circular, 1844. (DU Espy).

Espy's rainmaking ideas sometimes earned him derision, and other ideas of his may not have been taken so seriously in the scientific community had he not been in cooperation with scientists like Henry and Bache. In 1842, Espy wrote to Henry about a plan to have simultaneous observations of the weather made over as much territory as possible. Espy hoped that numerous observers could be found who would be willing to take part in such a major project "as soon as it is known, that their labors are not likely to be in vain, as thousands of insulated observations have heretofore been." Espy's purpose was specifically directed at discovering the shape, size and direction of storms. He implored Henry to assist him in setting up this project, which he was determined to see through with or without government aid.[52]

As it turned out, Espy did find employment under the surgeon general's office, working as a clerk collecting the meteorological observations ordered by the army at its posts. These observations gained the patronage of the Whig-controlled Senate in 1843 when it voted to supply $2,000 to continue the series of observations. Party discipline was back in force, with 96% of the Whigs backing the observations and 71% of Democrats opposed.[53] Throughout the 1840s, however, funding for these meteorological observations would become a political issue, and the fate of Espy's position from year to year depended on which party had the momentum in Congress.

In May 1846, funding for these observations, as well as for a state employed meteorologist, came under close scrutiny in the Democratically controlled House. Orville Hungerford (Democrat, NY) proposed an amendment to strike out funding for a meteorologist in the surgeon general's budget, arguing that the initial appropriation had been made illegally. John Quincy Adams supported Espy, believing that the public had benefited greatly from his observations, but Seaborn Jones (Democrat, GA) dismissed the idea of a meteorologist, commenting that Congress "need not pay him [Espy] for meteorological observations, or for making a storm which they could raise themselves as well as he could." After considerable debate in the committee of the whole, Hungerford's amendment was narrowly rejected 61 to 65.[54] But the issue was still not settled. A week later, George Jones (Democrat, TN) renewed the attack before the House itself, moving to "strike out so much as related to the Storm King." Jones's amendment to eliminate the funding for Espy (which was the same as Hungerford's from the previous week) now passed 92 to 77.[55]

52. Espy to Henry, July 2, 1842. (JHP V p.234).
53. U. S. *Senate Journal*. 1843. 27th Cong., 3rd sess., 10 February.
54. *Congressional Globe*. 29th Cong., 1st sess., 846. [May 20, 1846] Votes in the committee of the whole were not recorded in the journal.
55. *Congressional Globe*. 29th Cong., 1st sess., 876. [May 28, 1846] Eighty-six

Now 84% of Democrats voted to cut off funding while 90% of Whigs voted against. The Democrats were not yet willing to make such a commitment to science.

As a result, Henry and the Smithsonian Institution became involved in Espy's plan. In 1847 Henry asked Loomis to draw up a plan of meteorological observations that would allow the study of storms. Observers would be furnished with instruments by the Smithsonian Institution, and distributed at a rate of one for every 100 square miles. Henry was hopeful that this system of observations could be linked to a similar system in Canada, and suggested such a plan to Sabine.[56] Henry's overall plan embraced a series of meteorological observations to be made from the Arctic Circle to the Gulf of Mexico.[57] Sabine agreed, believing that in order to give full efficacy to Henry's plan, he should have the cooperation of British observers in Canada and the provinces to the north and west of Canada. He suggested that Henry ask the American government to apply to the British government through their ambassador, specifically citing "the example which has been so recently given by the Government of Great Britain in requesting the cooperation of other countries in magnetical and meteorological researches."[58] For his part, Sabine seemed sure that his government would take up the task.[59]

In 1848 the Smithsonian Institution appropriated $1,000 toward this system of observations, and Henry hoped that the British government would agree to participate in the wider project, once it saw the American system in operation.[60] By March 1848, Loomis also felt that even the American government might be willing to take up the project.[61] Funding for the proposed observations was included in an amendment offered by Samuel Vinton (Whig, OH) to the Naval Science Bill in the House of Representatives in June 1848. As before with the Ex Ex, debate in Congress broke between supporters of government expenditure on science and those who wished to limit the government's involvement altogether. Advocates of the observations were aided by the fact that control of the House was now in the hands of the (more science friendly) Whigs and their "progressive" Democratic allies. That same month, the House had voted to provide

---

Democrats and six Whigs voted to eliminate funding for Espy. Opposed were fifty-seven Whigs, sixteen Democrats, three American Republicans ("know-nothings") and one Independent. U. S. *House Journal.* 1846. 29th Cong., 1st sess., 28 May.
56. Henry to Sabine, August 13, 1847. (PRO BJ3/32/6–7).
57. Henry to Edwards, December 18, 1847. (JHP VII p.251).
58. Sabine to Henry, September 11, 1847. (PRO BJ3/30/75–76).
59. Sabine to Bache, December 1, 1847. (PRO BJ3/30/79).
60. Henry to Redfield, January 11, 1848. (JHP VII pp.259–260).
61. Loomis to Sabine, March 17, 1848. (PRO BJ3/25/166).

$5,000 for astronomic observations in Chile under the superintendence of Lieutenant Gillis.[62]

Vinton, a veteran of the Ex Ex debate twelve years earlier, introduced an amendment to support the observations along with letters of support from both Henry and Bache. Vinton stressed that the observations would benefit both science and navigation. He also praised Henry's idea of using telegraphs to speed up reporting times, allowing an advance warning of large storms that could save many lives, as well as helping commercial and naval vessels avoid rough weather. Alexander Stephens (Whig, GA) believed that "in no other way could the same amount of money be appropriated so as to be more useful in so serving the great objects of humanity." Charles Brown (Democrat, PA), a fan of Espy's lectures, praised the advances already derived from his theories. Brown cited examples of ships (including the navy ship *Hornet*) that had been lost, often with all hands aboard, because violent storms had blown in after they had left port. "Had they known what we now know," he claimed, perhaps they would have met a different fate.

Opponents of the observations (primarily *laissez-faire* Democrats) attacked the very propriety of state expenditure on science. Jacob Thompson (Democrat, MS) did not think it proper for Congress to make any appropriations for scientific experiments, whatever the potential gains. "No one contended that they should establish a professorship of chemistry, although many valuable improvements might be made."[63] Frederick Stanton (Democrat, TN) later rebuked this argument by invoking the example of Lewis and Clark. He pointed out that Jefferson himself had instructed Lewis and Clark to study the flora, fauna and climate of the region they explored as an example of state involvement with science. "What had this Government to do with the habits of reptiles, birds and insects?" he asked. "What had it to do with lightning, hail, snow ice, heat, cold and winds?"[64] But other Democrats were dismissive of the scientific goals of the project. Archibald Atkinson (Democrat, VA) "could not consent that the navy proper should be saddled with every scheme and humbug." George Jones (Democrat, TN) was harsher still, challenging the very legitimacy of the scientific goals. He attacked the proposal as a waste of public money on "visionary projects for the regulation of storms and the making of rain."[65] Indeed, Espy's support of rainmaking theories left him vulnerable to just such attacks.

However, in 1848 Congress did not dismiss Espy as a quack or as simply out to wrest funding from Congress for a harebrained "rainmaking" scheme. The limited debate and quick approval of the

62. U. S. *House Journal*. 1848. 30th Cong., 1st sess., 15 June.
63. *Congressional Globe*. 30th Cong., 1st sess., 826. [June 12, 1848]
64. Ibid., 845. [June 16, 1848]
65. Ibid., 826. [June 12, 1848]

observations were signs that American science had become more publicly respected and accepted in recent years. Similar requests had been made in the past, but had met with derision in the House. In 1846, Jones had sponsored the amendment that had killed funding for the observations. Now he was in the minority and science was treated with more consideration. The change in climate was apparent to many. During the debate, Stanton "congratulated the House on the progress which had been made in a due respect for science. That which had been laughed at two years ago now seemed to meet with general favor."[66]

Overall, Congress did seem more hospitable to scientific expenditure. The Whigs and "progressive" Democrats were more willing to invest in science and to involve the state with scientific enterprises, overriding the more cautious conservative Democrats. The "federal" ideas of the Whigs, more conducive to governmental support of science, now held sway. Joseph Root (Whig, OH) humorously commented on the political cost for the "stationary Democracy" which opposed such progress. Back in May 1846, conservative Democrats had voted down Espy's proposed salary as meteorologist. "What was the consequence?" joked Root. The elections in the fall of 1846 had been a Whig landslide, with the party picking up almost forty seats in the House. "*That* the stationary Democracy could understand; it was no theory."[67]

In the end, the amendment was easily approved and Congress appropriated $2,000 annually for Espy, who worked under the navy department.[68] Espy was to coordinate with the Smithsonian to ensure that they did not duplicate their efforts.[69] Because there was no recorded voice vote, it is not possible to analyze support for the observations by party or region. From the contents of the debate and the ease of the amendment's passage, though, it is clear that the bulk of the Whigs (both northern and southern) and the "progressive" Democrats backed the project.[70] Once again, as with the Ex Ex, opposition primarily came from conservative southern Democrats. There is no doubt that the attempts by American scientists to professionalize their field in the 1840s were paying off. Attempts to portray Espy as crank and his science as "rainmaking" were brushed aside, in part due to Henry and Bache's backing of the project. Rather than a "charlatan," Espy was seen as a "real" scientist. In 1850, the Democratically controlled

---

66. Ibid., 826. [June 12, 1848]
67. Ibid., 827. [June 12, 1848]
68. This amount included $1,500 as a salary and $500 for other expenses such as equipment and printing costs.
69. Henry to Dobbin, September 10, 1853. (JHP VIII pp.477–478). Ironically Dobbin, as secretary of the navy under Franklin Pierce, was now administering the same meteorological observations that he had voted against in the House in 1846.
70. *Congressional Globe*. 30th Cong., 1st sess., 843. [June 15, 1848]

Senate voted to publish 3000 additional copies of Espy's recent report
to the secretary of the navy on meteorology.[71]

Espy and Henry worked together on a general circular that appealed
for observers to volunteer their time for the new meteorological project. It went out November 1, 1848, and quickly generated a number of
positive answers. By 1852, more than 200 observers were taking part;
by 1854 there were 350.[72] The Smithsonian's system was supple-
mented by various independent systems set up in the states. New York
continued to conduct its series of meteorological observations, appro-
priating $1,500 a year to continue the observations into the 1850s. In
1849 the Smithsonian assumed the responsibility of buying instruments for New York.[73] The New York observations, along with those
made in Massachusetts and at various army posts, eventually were integrated into a single network directed by the Smithsonian.[74]

Plans for British cooperation continued to develop, with Henry assuring Sabine that the United States had not abandoned the project in
1848.[75] Henry met with Lefroy in Toronto in the fall of 1849 concerning cooperation between the observers in the British provinces and the
American system of meteorology.[76] Sabine however, warned Lefroy not
to begin work with Henry until the American application to the British
government had been officially sanctioned.[77] A way to increase the
number of observing stations by equipping all telegraph stations with
suitable instruments was suggested in 1849 to Henry.[78] The
Smithsonian had already embraced the idea of using telegraphic reports of meteorological observations. Bache saw the use of such up-to-
date reports, asking Henry to let him know if storms were approaching
his coastal surveying parties.[79] By 1852, the Smithsonian's system of
observations had extended to include Central American and Caribbean
stations as well.[80] Still the most important aspect of the project, in
Henry's opinion, was the cooperation between American and British
observers. "Nothing, in my opinion, can be more conducive to the

71. U. S. *Senate Journal.* 1850. 31st Cong., 1st sess., 20 July. The motion passed
    25 to 20. While a majority of Democrats opposed the expenditure, 46%
    supported the publication, along with 76% of the Whigs. Both Free Soil
    Senators opposed the measure.
72. Bruce, 194; Henry to Bache, May 6, 1854. (JHP IX p.68).
73. Beck to Henry, May 10, 1849. (JHP VII p.528).
74. Henry to Guyot, March 26, 1851. (JHP VIII pp.164–165).
75. Henry to Sabine, October 13, 1848. (PRO BJ3/49).
76. Henry to Baird, November 1, 1849 (JHP VII p.618).
77. Sabine to Lefroy, October 26, 1849. (PRO BJ3/40/121).
78. Jones to Henry, July 21, 1849. (JHP VII pp.575–576). Lamont suggested a
    similar system for Bavaria in 1852. Lamont to Sabine, August 13, 1852.
    (PRO BJ3/4/74).
79. Bache to Henry, September 27, 1850. (JHP VIII p.109).
80. Henry to Gray, November 6, 1852. (JHP VIII p.402).

promotion of knowledge and philanthropy among the nations of the Earth than a general system of this kind."[81]

In addition to the land-based observations that were envisioned across North America, the British government applied to the American state for a system of meteorological observations at sea. In December 1851, the secretary of the navy authorized Matthew Maury to begin putting together the American side of this project.[82] Maury of course turned to Henry for assistance in designing such a system.[83] Like Henry and Espy, Maury was concerned with the problem of storms, except that he was working from the viewpoint of navigation. Maury's approach to science was positively Herschelian. Like Herschel, Maury believed that data collecting could be easily parceled out to numerous observers. "Every one who observes wind and weather, and who is in the habit of noting the thermometer and the barometer, is already an observer whose services it is desirable to secure." Like Herschel, Maury saw missionaries as prime candidates for observers. Maury especially saw sailors as an "immense corps of laborers who are already in the field," and ships as "floating observatories." He also saw the value of simultaneous observations taken at multiple points. The very vastness of the atmosphere required "that its operations should be observed in all its parts, and watched from all points at the same time." Maury appealed to shipowners and masters "in all maritime countries" to participate in this project.[84] Maury also organized an international meteorological conference, held at Brussels in 1853.[85]

Maury's project soon became part of the overall North American plan set up by Henry and Sabine. "You should take charge of the dominion of the sea," wrote Henry to Maury, while "the Smithsonian Institution should collect observations from the eastern part of the United States, the Surgeon General of the Army from the Western, and the British Government from the northern part of the continent."[86] The Royal Society embraced the proposal on the grounds that accurate knowledge of meteorology and climate would benefit the "welfare and material interests of the people of every country." Writing to Maury, Christie pointed out that due to its fleet, Britain possessed a greater interest in the results than any other nation, and proposed to bring Maury's plan to the attention of the admiralty.[87] In his 1852 address to the British Association, Sabine highlighted the need for "the united

81. Henry to Sabine, May 7, 1852. (PRO BJ3/49).
82. Maury to Secretary of the Navy, November 6, 1852. Reprinted in Matthew Maury, *Explanations and Sailing Directions to Accompany the Wind and Current Charts* (Washington, DC: Harris, 1858), 35.
83. Maury to Henry, January 14, 1852. (JHP VIII p.279).
84. Maury, 29–31.
85. Bruce, 181.
86. Henry to Maury, May 31, 1852. (JHP VIII p.327).
87. Christie to Maury, May 1852. Reprinted in Maury, 32–35.

labours of the two greatest naval and commercial nations of the world
[to] be combined."[88] The British admiralty eventually ordered that meteorological observations be carried out on all ships.[89] Bache also realized the advantages British aid might bring to sea observations. In
1853, his interest in the course of the Gulf Stream led to the suggestion that the American and British governments work together to
chart its full extent.[90] In 1856 Maury worked with Senator James
Harlan (Republican, IA) to introduce a bill (S. 481) to create a national
weather service.

There were internal problems with this great alliance, caused by personal animosities. Although at one point Henry and Maury worked
closely, their professional relationship soured in later years. When
Maury's national weather service bill was introduced in December
1856 from the committee on agriculture, it sought $20,000 to extend
the existing system of observations (under the Smithsonian) and place
it under the superintendent of the National Observatory and
Hydrographic Office. Henry accused Maury to trying to set up his own
system of meteorological observations that would report to the surgeon general's office rather than through the Smithsonian. Henry saw
the bill as a threat to the Smithsonian's position and helped to defeat
it. Henry also worked to prevent Maury's election as president of the
American Association, which Henry felt would be the "triumph of
charlatanism and the disgrace of true science in our country."[91] At the
same time, Thomas Lawson, the surgeon general, resented the
Smithsonian taking the credit for a system of observations that he felt
were primarily being made at army posts and were reported to his office. In 1856, he pointed out to Henry that the army system had been
in operation for thirty-five years and that it accounted for all observations taken west of 95 degrees of longitude: "Thus much to show that
this Office did not need co-operation with anyone."[92]

ANGLO-AMERICAN COOPERATION—THE TORONTO OBSERVATORY

While American and British scientists continued to plan their
joint meteorological observations, another issue became an important point of contact for them. The Toronto observatory, part of the
colonial network of stations founded by the crusade, had been a major
base of operations for British geophysical research in North America.

---

88. Edward Sabine, "Address." *BAAS Report* 22 (1852), lix.
89. Sabine to Quetelet, October 23, 1853. (APS HS #11:1).
90. Henry to Sabine, June 1, 1853. (JHP VIII p.445).
91. Henry to Bache, July 31–August 4, 1855. (JHP IX p.273).
92. Lawson to Henry, April 5, 1856. (JHP IX pp.339–340).

Of the three colonial observatories manned by artillery officers (Hobart was a navy station), only Toronto continued to operate as an independent establishment into the 1850s.[93] However, after renewing the funding for the Toronto observatory three times (in 1842, 1845 and 1848) the British government was unwilling to continue its operation beyond 1851. Sabine was eager to keep the Toronto observatory functioning because he did not feel that enough time had elapsed to determine the annual magnetic variation. Additionally, given the plans for cooperation with the United States in the geophysical field, he wanted the Toronto observatory to serve as a focal point for the northern part of the project and hoped he could employ American assistance to keep it running.

Sabine made his appeal to Henry in the spring of 1850, stating that an application from a scientific institution in the United States to the Royal Society might encourage another renewal for the Toronto observatory. He suggested the American Academy in Boston, the American Philosophical Society or the Smithsonian as appropriate institutions. Sabine believed that the new self-registering instruments being tried out at Toronto could become a model for future American observatories, and suggested that Henry emphasize the Toronto observatory's connection to American magnetism and meteorology. Despite the considerable amount of coaching that Sabine gave Henry on this matter, he was eager that the American appeal appear objective. "It would no doubt greatly strengthen the hands of those who are engaged here in the promotion of these branches of Science," he commented. "Of course, any proposition coming from your side [of] the water must not appear to have originated in suggestions from this country."[94]

The continuation of the Toronto observatory became an important front in the joint Anglo-American scientific efforts. Without a base to anchor the Canadian observations, the overall North American geophysical plan was untenable. The work done at Toronto, declared Henry, was of "the highest importance to the advance of our knowledge of magnetism and meteorology."[95] What followed was a new lobbying effort, one designed to use American and British pressure to achieve the results desired from the British government. Despite its appearances, this new lobby was largely the work of Sabine and Henry, who worked closely together, often planning their moves in advance.

Just as the British had done in 1838, the Americans now turned to their counterpart to the British Association, by employing the newly

93. The instruments of the Cape magnetic observatory were to be transferred to the Cape astronomical observatory in 1845, while the St. Helena observatory had ceased operations in the spring of 1849. Bynam to Sabine, January 26, 1849. (PRO BJ3/27/386).

94. Sabine to Henry, May 17, 1850. (PRO BJ3/30/112).

95. Henry to Sabine, November 26, 1850. (PRO BJ3/49).

formed American Association for the Advancement of Science. Founded in 1848, the AAAS was patterned after the British Association, and vowed to "promote intercourse between those who are cultivating science in different parts of the United States; to give a stronger and more general impulse, and a more systematic direction to scientific research in our country; and to procure for the labours of scientific men, increased facilities and a wider usefulness."[96] At its meeting in August 1850, resolutions were adopted to continue the Toronto observatory, as well as to invite the British government and Hudson's Bay Company to participate in the meteorological system set up by the Smithsonian.[97] These resolutions were transmitted through the American minister at the court of St. James to the Prime Minister, Lord Palmerston.[98]

Sabine now went into action, tracking the American resolutions through the labyrinth of British government and prodding Lord Grey at the colonial office to take notice of them. Sabine realized that in the absence of Lord Rosse, President of the Royal Society (in Ireland until Easter), the resolutions would be referred to the treasurer of the Royal Society (Sabine himself!) for a report. Sabine thus asked Henry to furnish him with specific information as to the particular objects of the meteorological inquiry being conducted by the Smithsonian Institution and also what definite steps Henry wished the government and the Hudson's Bay Company to undertake in relation to it. "Perhaps it will be well that our demand should state to the *fullest extent* what you would wish us to do: the consideration on our part would then be much simplified." Sabine, realizing that he himself would be the government's prime resource on the matter, was preparing his response to the American resolutions (with additional material from Henry) even before the government asked for it![99]

Yet despite Henry and Sabine's joint efforts, the British government refused to continue operating the Toronto observatory due to the expense. With its demise, only the Rossbank observatory in Tasmania remained out of the four set up by the crusade. Rossbank itself fell victim to budget cuts in 1853, the Crimean War proving costly for the British government, resulting in reduced scientific expenditures. Just as the long period of relative European peace in the early nineteenth century gave the crusade a chance to gain state aid, so the return of

---

96. Bruce, 255–256. Bache later expressed a preference for going through the British and American Associations on scientific matters, rather than appealing directly to the governments of the two nations, as he feared that approach would bring "the politicians in between the savants." Bache to Henry, May 30, 1853. (SIA RU:7001).
97. Loomis to Lefroy, August 29, 1850. (PRO BJ3/25/192); Sabine to Rosse, October 22, 1850. (PRO BJ3/30/118).
98. Henry to Sabine, November 26, 1850. (PRO BJ3/49).
99. Sabine to Henry, December 26, 1850. (PRO BJ3/30/119–120).

war signaled the end of its objectives. Lieutenant Kay left Hobart in March 1853. Despite attempts by the local colonial government to keep the observatory running, it closed completely by the end of 1854.[100]

The American scientific community regretted the decision to discontinue the observations at Toronto. The regents of the Smithsonian appealed on "behalf of the science of our day," for the continuation of the observations at Toronto.[101] Henry was also disappointed by the loss, especially as the Smithsonian was in the process of founding its own observatory in Washington, DC to make corresponding observations with Toronto. He hoped that some last minute arrangement could be made to preserve the observatory.[102] Such a solution was finally found, not by the British or American governments, but by the government of Upper Canada.[103] In the summer of 1853, the provincial parliament appropriated £2,000 to repair the Toronto observatory and bring it back into operation.[104] In 1855, a new stone observatory designed by Frederick Cumberland was erected on the campus of the university to replace the original wooden cabin that had stood since 1840.[105] This observatory continued to operate throughout the nineteenth century. In 1856, Henry wrote to its new director George Kingston, offering to continue the cooperative efforts between them.[106]

Unfortunately, the plans for joint British-American observations across North America also fell victim to a lack of funding. Henry and Sabine were unable to carry out their (perhaps too) ambitious goal. However, the Smithsonian threw itself into the task of studying American meteorology.[107] In 1855 the Smithsonian entered into a partnership with the Patent Office to print and distribute blank forms to the observers.[108] Henry worked to find more volunteer observers,

100. The stone observatory buildings of Rossbank still stand on the grounds of the Royal Botanical Gardens and Government House in Hobart, now restored and used as residences. Ann Savours and Anita McConnell, "The History of the Rossbank Observatory, Tasmania." *Annals of Science*, 39:6 (November 1982), 541–543, 546.
101. Henry to Sabine, February 4, 1853. (PRO BJ3/49).
102. Henry to Sabine, February 9, 1853. (JHP VIII p.426).
103. Since the Durham report in 1839, the provincial governments of Canada had been given more autonomy.
104. Sabine to Cherriman, August 11, 1853. (PRO BJ3/40/164).
105. See cover illustration.
106. Henry to Kingston, December 6, 1856. (JHP IX p.422). The structure was disassembled in 1907 and relocated, becoming the home of the Department of Surveying and Geodesy for the University of Toronto. Since 1953, it has housed the offices of the Students' Administrative Council. The stone observatory still stands today, the oldest building on campus.
107. Henry to Lefroy, November 9, 1853. (JHP VIII pp.488–489).
108. Henry to Coffin, November 29, 1855. (JHP IX p.295).

instructing Charles Mason at the Patent Office to send out blanks to anyone who requested them.[109] Henry was drawn into the project of finding solutions to the problems of meteorology, admitting in 1858 that "I have been obliged myself to give for the first time in my life equal attention to meteorology."[110] Daily telegraphic reports from throughout the country flowed into the Smithsonian, allowing Henry to set up a meteorological map depicting clear, cloudy, rainy and snowy weather at the Smithsonian that became "a source of great interest to the visitors of the Institution."[111] The foundations of American meteorology had been laid.

▓ NEW FRONTIERS

Sabine hoped that the long series of magnetic observations taken at the colonial observatories in the 1840s might allow a concise solution of the whole geomagnetic problem. Accordingly, in 1850 he had written to Michael Faraday, asking him to lend his talents to the task. Sabine believed that Faraday's work in magnetism could allow him to solve the problem of secular variations by assuming that the Earth itself was a giant magnet. Gauss had already taken a step in this direction by arguing that the causes of geomagnetism lay within the Earth (although Sabine was quick to point out that "this does not apply to the periodical variations, or to the disturbances, which may all be external notwithstanding"). The search for a physical solution to the geomagnetic phenomena which the colonial observatories recorded was a daunting one, but one that offered a promise to be "the *opus magnus* [sic] after Newton's explanation of Gravitation."[112]

Faraday was a fan of Sabine's work, declaring that "his investigations [were] of the utmost value to science."[113] Like Gauss, Faraday held that "the *seat* of the terrestrial magnetic force" was in the Earth (although he allowed that the Sun, through its daily motion, could be the source of some variations in that force). He even suggested that the daily heating and cooling of oxygen in the atmosphere might have a magnetic influence.[114] But the Sun could only be a small part of the whole effect.[115] Despite the years of data, Faraday concluded that until Sabine could supply "comparisons & collections" of the

109. Henry to Mason, January 31, 1856. (JHP IX p.312).
110. Henry to Herschel, July 22, 1858. (APS HS #1:5).
111. Henry to Sabine, July 8, 1861. (APS HS #1:6).
112. Sabine to Faraday, September 26, 1850. (PRO BJ3/30/116–118).
113. Faraday to Bell, May 15, 1851. (FC IV p.294).
114. Faraday to Whewell, June 8, 1852. (FC IV p.395); Faraday to Sabine, September 17, 1850. (FC IV p.185).
115. Faraday to Sabine, January 7, 1853. (FC IV p.467).

variations, he was "working without the necessary data that are within our reach."[116]

However, the results of the observations did contribute to a new field of science: solar-terrestrial physics. Ever since Galileo, astronomers had been studying sunspots. In 1844, the German astronomer Heinrich Schwabe suggested that the appearance of sunspots followed a regular eleven-year cycle from maximum to maximum.[117] Alexander von Humboldt found Schwabe's theory appealing, and published it as part of his mammoth work *Kosmos*. Sabine wondered if the sunspot cycle had any noticeable terrestrial effects. Like Hansteen and von Humboldt, Sabine held a unified view of nature and believed that cosmic forces could influence terrestrial magnetism. So he began to look back over the observations from the colonial observatories.

In the early days of observing, the various stations had recorded great magnetic storms, such as the one that had attracted so much attention in 1841, which occasionally swept over the Earth. Indeed, establishing that these storms affected the entire surface of the globe simultaneously was one of the earliest results of Herschel's system. In 1851, Sabine examined the observations from the Toronto and Hobart observatories for the years 1842–1848. These two observatories he found ideally placed for contrasting the results of the years of observations. They were at nearly the same latitude, but in opposite hemispheres that gave them opposing seasons. Their longitudinal difference meant that observations made during the day at one observatory occurred during the night at the other. He had realized that the evidence for the simultaneity of the storms was "far too systematic" to be an accident.[118] In 1852, Lamont reported that the occurrence of the magnetic storms seemed to have a regular period of ten years.[119] Sabine now sought a common, extraterrestrial cause for these storms that would explain both their omnipresence and their periodicity.

In 1852, Sabine began to compare the pattern of magnetic disturbances with variation in the frequency and magnitude of sunspots described by Schwabe. At the Toronto and Hobart observatories, he found that the variations caused by the disturbances corresponded, suggesting a common cause was in effect. He discovered that the storms also followed a pattern of minimum and maximum and that it was the same cycle as the sunspots. Sabine suggested that the sunspot cycle

---

116. Faraday to Sabine, January 12, 1853. (FC IV p.472).
117. Samuel Heinrich Schwabe, "Sonnen-Beobachtungen im Jahre 1843," *Astronomische Nachrichten*, 21 (1844) 233–236.
118. Edward Sabine, "On Periodical Laws Discoverable in the Mean Effects of the Larger Magnetic Disturbances." *Philosophical Transactions*, 141 (1851), 123–139.
119. Johann Lamont, "On the Ten-Year Period Which Exhibits Itself in the Diurnal Motion of the Magnetic Needle." *Philosophical Magazine*, 3 (1852), 428–435.

itself was directly affecting geomagnetism, and that influence manifested itself as magnetic storms. "It is certainly a most striking coincidence, that the period, and the epochs of minima and maxima, which M. Schwabe has assigned to the variation of the solar spots, are absolutely identical with those which have been here assigned to the magnetic variations." Sabine's ideas about cosmic forces influencing the Earth seemed to have been justified. "The sun must be recognized as at least the primary source of all magnetic variations which conform to a law of local hours."[120] Herschel's global system of induction now expanded into a cosmic system. Sabine spoke of a "cosmical connexion" and declared the Sun to be the "primary source" of the terrestrial magnetic variations.[121] The cause of the great magnetic storm of 1841 had been found and the new field of solar-terrestrial physics opened for exploration.

Sabine's discovery caused a stir in the field of geomagnetism. John Welsh at the Kew Observatory remarked on the coincidence of the magnetic storms and solar spots after reading an early copy of Sabine's monograph. "If it is an accident it is a very singular one and must by no means be allowed to rest."[122] George Bond wrote to Bache, urging him to press Henry to investigate Sabine's findings.[123] After the publication of Sabine's paper, Herschel declared, "we stand on the verge of a vast cosmical discovery such as nothing hitherto imagined can compare."[124] Faraday invited Sabine to give a lecture on the topic at the Royal Institution.[125] The discovery also set off a brief dispute over who had made the connection first. In August 1852, Johann Rudolf Wolf wrote to Faraday claiming to have detected a 10.33 year period in the cycle of sunspots and magnetic disturbances from the works of Schwabe and Lamont.[126] In a later paper, he argued that a period of 11.11 years perfectly fit both the sunspot cycle and the period of magnetic storms.[127] Wolf claimed a nearly simultaneous publication of the discovery along with Sabine. However, Sabine later claimed that Wolf had overrated his achievement by dating Sabine's own discovery several months later than it had actually occurred.[128] Sabine had already declared that Wolf had been "forestalled by Lamont in the period of the variations" of declination and by Sabine himself "in the period of the supposed irregular

120. Edward Sabine, "On Periodical Laws Discoverable in the Mean Effects of the Larger Magnetic Disturbances, No. II." *Philosophical Transactions*, 142 (1852), 103–124.
121. Sabine, *Report* 1852, lv, liii.
122. Welsh to Sabine, April 23, 1852. (PRO BJ3/32/67–68).
123. Bond to Bache, July 7, 1852. (SIA RU:7053).
124. Herschel to Faraday, November 10, 1852. (FC IV p.443).
125. Faraday to Sabine, November 13, 1852. (FC IV p.447).
126. Wolf to Faraday, August 2, 1852. (FC IV p.410).
127. Wolf to Faraday, November 4, 1852. (FC IV p.438).
128. Sabine to Faraday, January 4, 1853. (FC IV p.465).

disturbances or storms—and in the coincidence of these periods with that of the solar spots."[129] He saw no reason to share the credit.

Other terrestrial phenomena were also linked to solar activity. In 1859, the first evidence linking solar flares to magnetic disturbances was recorded at the Kew Observatory outside London. On August 28 of same year, a massive aurora borealis display began in the northern hemisphere, lasting until September 4. Observers from Canada to Cuba and from Europe to California recorded its appearance and effects. Several days after the aurora first appeared, "extraordinarily great" magnetic disturbances were also recorded.[130] This display helped to confirm the long established link between aurora and magnetic storms (Henry had reported a possible link between an aurora and a magnetic disturbance as early as 1832). Eventually, the causes of aurora were also linked to solar activity.

Sabine stated the obvious when he gave credit to the British colonial observatories for discovering the link between Schwabe's observations of sunspots and the magnetic disturbances.[131] The contribution of Herschel's system of observing stations had been vital. Without an extended series of continuous observations from places as distant as Canada and Tasmania, Sabine's conclusion that the magnetic storms were being caused by solar activity could not have been so readily arrived at or accepted. Now what was needed were more years of observations to confirm the pattern. The continuing observations at Toronto helped that cause. The Magnetic Crusade had again delivered promising results that inspired new research. Astronomy, meteorology and geomagnetism could all be studied at once.[132] For some, Sabine's discovery of solar-terrestrial physics was enough to justify the expense and efforts of the whole global project. In 1855, George Bond commented to Sabine that the "single discovery of the connection between the solar spots and the terrestrial magnetic disturbances is of itself of an importance sufficient to repay the outlay of time & labor which have been required in conducting these establishments."[133]

---

129. Sabine to Faraday, November 11, 1852. (FC IV p.444).
130. Elias Loomis, "The Great Aurora Exhibition." *American Journal of Science and Arts.* 28 (November 1859), 385–389.
131. Sabine, *Phil. Trans.* 1852, 122.
132. Sabine to Hansteen, January 7, 1860. (ITA).
133. Bond to Sabine, March 19, 1855. (APS HS #1:6).

Conclusion

The results of the Magnetic Crusade and the geophysical system of observations it inspired were contingent upon a confluence of factors in science and state during the early nineteenth century. Necessary preconditions for the enterprise had to be combined with individual actors who were well placed in both scientific and political fields to fulfill their role in the venture. The success of the crusade depended upon the ability of these actors to integrate imperial expansion and Baconian science. The final outcome required the successful cooperation of multiple elements of both science and empire.

The system of observations that grew out of the initial crusade was a major achievement in science. But how was it different from what had gone before? Measurement of geophysical phenomena was an old story by the early nineteenth century. From South America to Siberia, from Humboldt to Hansteen, countless measurements of magnetic variations and meteorological events had been made. Physical observatories were also nothing new. By the nineteenth century, the German *Verein* comprised a series of observatories that stretched across Europe, and Humboldt's 1836 letter had already suggested colonial observatories for Britain. Finally, cooperation between science and state was also not unprecedented. In the eighteenth century, the British government had invested in a voyage to observe the transits of Venus as well as supporting Cook's explorations in the Pacific. Aside from its ambitious goals, how did the system launched in 1839 by the crusade differ from those projects attempted before? And what about its long-term consequences?

In the first case, the system engendered by the crusade was quantitatively beyond anything that then existed or had been attempted. Observatories in British colonial outposts covered much of the world, in both hemispheres. By contrast, even the German *Verein* covered only part of Europe, a decidedly local area of the globe. The distance between the Cape observatory and Rossbank at Hobart far exceeded any scope the *Verein* possessed. This coverage was important because of the particular nature of the geophysical sciences being studied.

The development of geophysical sciences in this period occurred in tandem with the growth of empire. Whether through the establishment of colonies by the British and French governments or the continental expansion of the United States and Russia, Western science reached new regions and found new ways of studying science on a global scale. Indeed, the central prerequisite for the study of new geophysical sciences was this very expansion, which brought with it the ability to study phenomena such as the terrestrial magnetic field or meteorology, topics which could not be seen completely except through a wide lens. Only through observations conducted at distant points could scientists even begin to grasp the nature of the geomagnetic and meteorological sciences; local European observations were not enough. A single observatory in London could not detect the worldwide effects of a magnetic storm or study a weather front hundreds of miles away. Only by globalizing induction could global sciences be studied.

Scientific research on such a worldwide scale by nature was inductive, requiring a commitment to science best exemplified by Humboldt in the early years of the nineteenth century. Later scientists followed in his path and applied similar methods to the study of those fields of geophysics that can still properly be called Humboldtian. The chief protagonists in the events of the Magnetic Crusade were Herschel and Sabine. Both adopted Humboldtian-style science, although with their own modifications. Both were well placed in both scientific and political society to see through their goals.

A second difference is qualitative. The system of observations set up by the crusade provided a continuous, regular collection of data over many years. Earlier attempts to collect data across hundreds of miles (such as Hansteen's trek across Siberia) had brought back mountains of data, little of which could be used because there was nothing with which to compare it. It was not just observations at distant points that were needed, but *simultaneous* observations. Even the measurements of magnetic variation taken onboard ships were of little use for the construction of general theories unless there was another set of data taken at the same time elsewhere. Temporary surveys only indicated physical phenomena at one time and place. It was difficult to extrapolate conditions at any other time or place apart from those where the observations had originally been conducted.

After the crusade, a system was in place that would not just cover the globe geographically, but one that would conduct series of observations that lasted for years. Again, given the nature of the sciences being studied, this was a necessity (secular variations on the magnetic field could take years alone just to detect). Unlike Cook's voyages (or even the *Beagle's*), the system of observations founded by the crusade was not just a one-time investment, but one that continued through various renewals for many years, vastly increasing its usefulness and results. Here the British political system clearly favored state sponsored science, allowing observations to continue even through the transition of ministries, rather than facing a new vote for approval in Congress after every election.

Finally, the crusade set a new model for cooperation between science and state. In both the extent and expense of the commitment, the British government set a precedent for the future of "big" science. In many ways, the success of the crusade depended upon timing. First, the expanding British Empire allowed science to observe in new locations and gave the state reason to look to science for advice. Geophysical sciences then reinforced imperial expansion. The study of terrestrial magnetism contributed to surveying methods and suggested possibilities that new navigational techniques could be achieved through the study of magnetic variation. The study of climate and meteorology provided information on which crops could be profitably planted in new lands, while geological surveys found new mineral sources to be exploited. Second, the recent creation of the British Association also provided a platform to lobby for government support. Perhaps there is no clearer contrast between British and American efforts in this period. British scientists benefited greatly from their own organization and the connections that men of science enjoyed with men of politics. In the United States, the fledgling scientific societies were less influential and supporters of scientific proposals unable to propel them through Congress without significant compromise. Finally, the presence of important figures like Herschel and Sabine, well positioned to drive the project to completion, was also essential.

Herschel stands out as the primary early mover of the magnetic project (that the start of the successful lobby for the crusade began only months after his return from Africa must not be overlooked). His inductive philosophy, combined with his own colonial experiences at the Cape convinced him of the potential of converting imperial science to global science. An aristocratic scientist who had inherited both scientific eminence and political position from his father, Herschel was able to move between the worlds of science and politics in a way that foreshadowed the later combined efforts of science and state in the crusade. Herschel's position was both important and necessary to the success of the project. Without his urging, the plan for colonial observatories might have been dropped and the crusade converted into yet another one-time expedition. His efforts shaped the form of the Magnetic Crusade, fulfilling the hopes for universal science that he had been nursing for years. The colonial observatories founded by the 1839 expedition were the direct descendents of the scientific stations he had called for ten years before in the *Preliminary Discourse*. In

Herschel we find all of the elements that made the venture a success: inductive science, political influence and imperial connections.

The influence of Sabine, Herschel's partner and successor in the project, cannot be ignored either. Within Sabine were elements that brought the project to its successful conclusion. A military man, Sabine provided the practical side of the venture. While Herschel was concerned with theories (and the occasional flight of fancy), Sabine was anchored in the real world, indefatigably carrying out the everyday work of collecting, analyzing and printing the results of Herschel's system of data collectors. Long after Herschel had moved on to other interests, Sabine continued to work on their project, making one of the most significant discoveries suggested by the data in his recognition of the connection between the cycle of solar spots and frequency of magnetic storms. While Herschel had constructed the instrument, Sabine was its most proficient operator.

Even with the resources of the British empire, the magnetic project was unable to go it alone. Here the universal aspect of Herschel's science can truly be seen. His system of geophysical observations was connected not just through colonial outposts, but also with those of other nations. The British system became part of an international, cooperative venture that literally spanned the globe. Whewell later declared that such a scheme, "combining worldwide extent with the singleness of action of an individual mind," was unprecedented.[1] His words recall Herschel's own definition of science: the knowledge of many attainable by one. The single mind responsible for this scheme was clearly Herschel's.

The descendents of Herschel's system of observations are still with us today, doing everything from studying periodic reverses in the Earth's magnetic field to forecasting tomorrow's weather. In the United States, meteorological observations directed back to the Smithsonian once allowed for the prediction of storms. Now the study of sunspots and their effect on the magnetic field of Earth allows us to predict communications outages. This global (or to use the nineteenth century term, "cosmic") view of the world became more apparent in the twentieth century. The realization that science can be nonlocal has contributed to new views on the environment and humanity's place in (and impact upon) the world. From solar flares to hurricanes, every time science studies geophysical phenomena, it relies in part upon that observational system created in 1839. For Herschel and his ideas of universal science, there could be no greater achievement.

---

1.   Whewell, 1857, III:51.

Works Consulted

The materials used in this project divide nicely between political and scientific sources. Records of the Board of Longitude are kept at Cambridge along with the papers of the Royal Greenwich Observatory. Both are also available on microfilm at the Public Record Office (now the National Archives). Debates in Congress are summarized in publications such as the *Congressional Globe*, published for public sale. Official accounts in the House and Senate *Journals* are limited in this period, containing only basic information on bills and votes with no attempt to include debates.

On the scientific side, sources tended to be of two types: published papers and private correspondence. British papers aimed at a scientific audience were often published in the journals of the two major societies, the *Philosophical Transactions* of the Royal Society and the *Report* of the British Association for the Advancement of Science. Articles aimed at a more general (but still educated audience) were often found in the *Quarterly Review* or the various philosophical magazines of the period. Private correspondence of the scientists involved in the crusade is found in several British repositories. Fortunately, the bulk of Herschel's material has remained intact and was deposited at the Royal Society in the nineteenth century. These papers are also available on microfilm. Sabine's papers and correspondence are divided between the Royal Society and the Public Record Office. The largest collection of Herschel material outside of Britain is at the University of Texas, Austin. Various smaller numbers of Herschel and Sabine letters are in the collections of their correspondents: at Cambridge (Whewell), the British Library (Babbage), the Public Record Office (Ross), St Andrew's University (Forbes) and the Institute for Theoretical Astrophysics in Oslo (Hansteen).

The only major American scientific publication of the period was the *American Journal of Science and Arts* (often called *Silliman's* after its editor). The Smithsonian Institution has published Henry's correspondence, although these volumes represent only a small portion of the total. The Smithsonian archives also hold a sizeable collection of Bache's correspondence and papers. Finally, the American Philosophical Society Library in Philadelphia has a wide collection of materials relating to the history of science, including microfilms of some papers from European repositories.

I. MANUSCRIPT COLLECTIONS

Additional Manuscripts. British Library, London.

Bache Papers. Smithsonian Institution Archives, Washington, DC.

Darwin Papers. American Philosophical Society, Philadelphia.

East India Company Papers. American Philosophical Society, Philadelphia.

Elliot Papers. National Maritime Museum, London.

English Scientific Autographs. American Philosophical Society, Philadelphia.

Espy Papers. Duke University, Durham, NC.

Evans Papers. Duke University, Durham, NC.

Fitzroy Papers. Public Record Office, London.

Forbes Papers. St. Andrews University, St. Andrews.

Hansteen Papers. Institute for Theoretical Astrophysics, University of Oslo.

Henry Papers. Smithsonian Institution Archives, Washington, DC.

Herschel Family Papers. University of Texas, Austin, TX.

Herschel Papers. American Philosophical Society, Philadelphia.

Herschel Papers. Royal Society, London.

History of Science Papers. American Philosophical Society, Philadelphia.

Kew Observatory Papers. Public Record Office, London.

Meteorological Department Papers. Public Record Office, London.

Minto Papers. National Library of Scotland, Edinburgh.

Miscellaneous Manuscripts. Royal Society, London.

Patterson Papers. Duke University, Durham, NC.

Ross Papers. Public Record Office, London.

Royal Greenwich Observatory Papers. Public Record Office, London.

Sabine Correspondence and Papers. Public Record Office, London.

Sabine Papers. Royal Society, London.

Terrestrial Magnetism Archive. Royal Society, London.

Wilkes Family Papers. Duke University, Durham, NC.

II. PUBLISHED PRIMARY SOURCES

Bond, Henry. *The Longitude Found* (London: Godbid, 1676).

"Celebration of the Hundredth Anniversary, May 25, 1843." *Proceedings of the American Philosophical Society* 3:27 (May 25–30, 1843): 1–36.

Christie, Samuel Hunter. "Discussion of the Magnetical Observations Made by Captain Back, R.N. during His late Arctic Expedition." *Philosophical Transactions* 126 (1836): 377–415.

_____. "Report on the State of our Knowledge Respecting the Magnetism of the Earth." *BAAS Report* 3 (1833): 105–130.

Espy, James. *The Philosophy of Storms* (Boston: Little and Brown, 1841).

Evans, David S., editor. *Herschel at the Cape* (Austin: University of Texas Press, 1969).

Farrar, John and Joseph Lovering. *Elements of Electricity, Magnetism and Electrodynamics* (Boston: Crocker & Ruggles, 1842).

Foster, Henry. "Corrections to the Reductions of Lieutenant Foster's Observations on Atmospherical Refractions at Port Bowen; with Addenda to the Tables of Magnetic Intensities at the Same Place." *Philosophical Transactions* 117 (1827): 122–125.

Gray, Jane, editor. *Letters of Asa Gray* (New York: Houghton, 1894). 2 volumes

Halley, Edmund. "A Theory of the Variation of the Magnetical Compass." *Philosophical Transactions* 13 (1683): 208–221.

_____. "An Account of the Cause of the Change of the Variation of the Magnetical Needle; With an Hypothesis of the Structure of the Internal Parts of the Earth: As It Was Proposed to the Royal Society in One of Their Late Meetings." *Philosophical Transactions* 16 (1686–1692): 563–578.

_____. *The Description and Uses of a New and Correct Sea Chart of the Whole World* (London: Hartigan, 1701).

Henry, Joseph. "On a Disturbance of the Earth's Magnetism." *American Journal of Science and Arts* 22 (April 1832): 143–155.

Herschel, John. "Address." *BAAS Report* 15 (1845): xxvii–xliv.

_____. "Humboldt's Kosmos." *Edinburgh Review* 87 (January 1848): 90–121.

_____. *Instructions for Making and Registering Meteorological Observations in Southern Africa* (London: Bradbury and Evans, 1838).

_____. "On a Remarkable Application of Cotes's Theorem." *Philosophical Transactions* 103 (1813): 8–26.

_____. *Preliminary Discourse on Natural Philosophy* (New York: Johnson Reprint Corporation, 1966).

_____. "Sound." *Encyclopedia Metropolitana* (London, 1845) 4:763–824.

_____. "Terrestrial Magnetism." *Quarterly Review* 66 (1840): 271–312.

_____. *A Treatise on Astronomy* (London: Longman, 1833).

_____. "Whewell on Inductive Sciences." *Quarterly Review* 66 (1840): 177–238.

_____ and Humphrey Lloyd. "Report on the subject of a series of Resolutions adopted by the British Association at their Meeting in August, 1838, at Newcastle." *BAAS Report* 9 (1839): 31–42.

Howgate, Henry. "Congress and the North Pole." *United Service: A Quarterly Review of Military and Naval Affairs* 2:1 (January 1880): 72–84.

Hume, David. *An Inquiry Concerning Human Understanding* (New York: Bobbs-Merrill, 1955).

"Instructions for the Scientific Expedition to the Antarctic Regions, prepared by the President and Council of the Royal Society." *London and Edinburgh Philosophical Magazine* 95 (September 1839): 177–193.

James, Frank A. J. L., editor. *The Correspondence of Michael Faraday* (London: Institution of Electrical Engineers, 1999). 4 volumes

"John Herschel Letter." *South African Libraries* 7 (1940): 138–154.

Lamont, Johann. "On the Ten-Year Period Which Exhibits Itself in the Diurnal Motion of the Magnetic Needle." *Philosophical Magazine* 3 (1852): 428–435.

Lefroy, John Henry. *Diary of a Magnetic Survey of the Dominion of Canada* (London: Longman, Green & Co., 1883).

Loomis, Elias. "On Two Storms Which Were Experienced throughout the United States." *Transactions of the American Philosophical Society* 9:2 (1845): 161–184.

_____. "The Great Aurora Exhibition." *American Journal of Science and Arts* 28:84 (November 1859): 385–389.

Maury, Matthew. *Explanations and Sailing Directions to Accompany the Wind and Current Charts* (Washington, DC: Harris, 1858).

Morrell, Jack and Arnold Thackray, editors. *Gentlemen of Science: Early Correspondence of the BAAS* (London: Royal Historical Society, 1984).

Murchinson, Roderick and Edward Sabine. "Address." *BAAS Report* 10 (1840): xxxiii–xlviii.

Peale, Titian Ramsey. "The South Sea Surveying and Exploring Expedition." *American Historical Review* 3:30 (June 1874): 244–251.

Reingold, Nathan, editor. *Science in Nineteenth Century America: A Documentary History* (Chicago: University of Chicago Press, 1964).

"Report of a Committee." *BAAS Report* 10 (1840): 427–431.

"Report of a Committee." *BAAS Report* 11 (1841): 38–41.

"Report of the Committee." *BAAS Report* 12 (1842): 1–11.

"Report of the Committee." *BAAS Report* 13 (1843): 54–60.

"Report of the Committee." *BAAS Report* 14 (1844): 143–153.

"Report of the Committee." *BAAS Report* 15 (1845): 1–73.

"Report of the Joint Committee." *BAAS Report* 28 (1858): 295–305.

Reynolds, Jeremiah N. *The South Sea Surveying and Exploring Expedition* (New York: Harper & Brothers, 1836).

Ross, James. "On the Position of the North Magnetic Pole." *Philosophical Transactions* 124 (1834): 47–52.

Rothenberg, Marc, editor. *The Papers of Joseph Henry* (Washington, DC: Smithsonian Institution Press, 1992–2002).

Sabine, Edward. "An Account of Experiments to Determine the Acceleration of the Pendulum in Different Latitudes." *Philosophical Transactions* 111 (1821): 163-190.

_____. "An Account of Experiments to Determine the Amount of the Dip." *Philosophical Transactions* 112 (1822): 1–21.

_____. "Contributions to Terrestrial Magnetism." *Philosophical Transactions* 130 (1840): 129–155.

_____. "Observations on the Magnetism of the Earth." *American Journal of Science and Arts* 17:1 (1830): 145–156.

_____. "On Irregularities Observed in the Direction of the Compass Needle." *Philosophical Transactions* 109 (1819): 112–122.

_____. "On Periodical Laws Discoverable in the Mean Effects of the Larger Magnetic Disturbances." *Philosophical Transactions* 141 (1851): 123–139.

_____. "On Periodical Laws Discoverable in the Mean Effects of the Larger Magnetic Disturbances. No. II" *Philosophical Transactions* 142 (1852): 103–124.

_____. "On some of the results obtained at the British Colonial magnetic Observatories." *BAAS Report* 24 (1854): 355–368.

_____. "On the Phenomena of Terrestrial Magnetism." *BAAS Report* 5 (1835): 61–90.

_____. "Report on the Variation of the Magnetic Intensity." *BAAS Report* 7 (1837): 1–37.

Schwabe, Samuel Heinrich. "Sonnen-Beobachtungen im Jahre 1843." *Astronomische Nachrichten* 21 (1844): 233–236.

Sixth Census or Enumeration of the Inhabitants of the United States (Washington, DC: Blair & Rives, 1841).

Todhunter, Isaac, editor. *William Whewell* (New York: Johnson Reprint Corporation, 1979).

von Humboldt, Alexander. "On the Advancement of the Knowledge of Terrestrial Magnetism." *London and Edinburgh Philosophical Magazine* 9 (1836): 42–53.

Whewell, William. *Astronomy and General Physics* (London: Pickering, 1834).

_____. "Herschel's Preliminary Discourse." *Quarterly Review* 45 (1831): 374–405.

_____. *History of the Inductive Sciences* (London: John Parker, 1857).

_____. *Philosophy of the Inductive Sciences* (London: John Parker, 1847).

_____. "Report on the Recent Progress and Present Condition of the Mathematical Theories of Electricity, Magnetism and Heat." *BAAS Report* 5 (1835): 1–33.

III. SECONDARY SOURCES

Agassi, Joseph. "Sir John Herschel's Philosophy of Success," *Historical Studies in the Physical Sciences* 1 (1969): 1–36.

Anderson, Katharine. *Predicting the Weather* (Chicago: University of Chicago Press, 2005).

Ashworth, William J. "The Calculating Eye: Baily, Herschel, Babbage and the Business of Astronomy." *British Journal for the History of Science* 27:4 (December 1994): 409–441.

_____. "John Herschel, George Airy and the Roaming Eye of the State." *History of Science* 36 (1998): 151–178.

Basalla, George. "The Spread of Western Science." *Science* 156 (May 1967): 611-622.

Bayly, Christopher. *Empire and Information* (Cambridge: University of Cambridge Press, 1997).

Becher, Harvey. "Radicals, Whigs and Conservatives: The Middle and Lower Classes in the Analytic Revolution at Cambridge in the Age of Aristocracy." *British Journal for the History of Science* 28:4 (December 1995): 405–426.

Brockway, Lucille. *Science and Colonial Expansion* (New York: Academic Press, 1979).

Bruce, Robert. *The Launching of Modern American Science* (New York: Knopf, 1987).

Buttman, Günther. *The Shadow of the Telescope* (New York: Charles Scribner's Sons, 1970).

Cain, Peter and Anthony Hopkins. *British Imperialism* (London: Longman, 1993).

Cannon, Susan Faye. *Science in Culture* (New York: Dawson, 1978).

Cannon, Walter. "John Herschel and the Idea of Science." *Journal of the History of Ideas* 22:2 (April–June 1961): 215–239.

Carter, Christopher. "Magnetic Compass," *Oxford Companion to World Exploration* (New York: Oxford University Press, 2007): 204–205.

Cawood, John. "Comments," *Human Implications of Scientific Advance* (Edinburgh: Edinburgh University Press, 1978): 139–149.

_____. "The Magnetic Crusade: Science and Politics in Early Victorian England." *Isis* 70:4 (December 1979): 493–518.

_____. "Terrestrial Magnetism and the Development of International Collaboration in the Early Nineteenth Century." *Annals of Science* 34:6 (November 1977): 551–587.

Chambers, David and Richard Gillespie. "Locality in the History of Science: Colonial Science, Technoscience, and Indigenous Knowledge. *Osiris* 15 (2001): 221–240.

Crowe, Michael with David Dyck and James Kevin, editors. *A Calendar of the Correspondence of Sir John Herschel* (Cambridge: Cambridge University Press, 1998).

Current, Richard. *American History* (New York: Alfred Knopf, 1964).

Drayton, Richard. *Nature's Government* (New Haven: Yale University Press, 2000).

Fogg, G. E. "The Royal Society and the Antarctic." *Notes and Records of the Royal Society of London* 54:1 (January 2000): 85–98.

Friendly, Alfred. *Beaufort of the Admiralty* (New York: Random House, 1977).

Gardner, Brian. *The East India Company* (New York: Dorset Press, 1971).

Garland, G. D. "The Contribution of Carl Friedrich Gauss to Geomagnetism." *Historia Mathematica* 6 (1979): 5–29.

Gascoigne, John. *Science in the Service of Empire* (Cambridge: Cambridge University Press, 1998).

Goetzmann, William. *New Lands, New Men* (New York: Viking, 1986).

Gooding, David. "'Magnetic Curves' and the Magnetic Field: Experiment and Representation in the History of a Theory," *The Uses of Experiment* (Cambridge: Cambridge University Press, 1989): 183–219.

Hall, Marie Boas. "The Distinguished Man of Science," *John Herschel 1792–1871: A Bicentennial Commemoration* (London: Royal Society, 1992): 115–124.

Hankins, Thomas. "A 'Large and Graceful Sinuosity': John Herschel's Graphical Method." *Isis* 97:4 (December 2006): 605–633.

Holt, Michael. *The Rise and Fall of the American Whig Party* (Oxford: Oxford University Press, 1999).

Jackson, C. Ian. "Exploration as Science: Charles Wilkes and the U.S. Exploring Expedition." *American Scientist* 73:5 (September–October 1985): 450–461.

Jackson, Mike. "In the Shadow of the Comet." *IRM Quarterly* 7:4 (Winter 1997–1998): 1, 5–7.

_____. "Von Humboldt's Equinoxial Journey." *IRM Quarterly* 11:2 (Summer 2001): 1, 9–10.

Jankovich, Vladimir. "Ideological Crests Versus Empirical Troughs: John Herschel's and William Radcliffe Birt's Research on Atmospheric Waves, 1843–50." *British Journal for the History of Science* 31:1 (March 1998): 21–40.

Jonkers, A. R. T. *Earth's Magnetism in the Age of Sail* (Baltimore: Johns Hopkins Press, 2003).

Josefowicz, Diane. "Experience, Pedagogy and the Study of Terrestrial Magnetism." *Perspectives on Science* 13:4 (Winter 2005): 452–494.

Keay, John. *The Great Arc* (New York: HarperCollins, 2000).

King-Hele, Desmond, editor. *John Herschel 1792–1871: A Bicentennial Commemoration* (London: Royal Society, 1992).

Latour, Bruno. *Science in Action* (Cambridge: Cambridge University Press, 1987).

MacKenzie, John. "Introduction," *Imperialism and the Natural World* (Manchester: Manchester University Press, 1990): 1–14.

MacLeod, Roy. "Introduction," *The Parliament of Science* (Northwood: Science Reviews Ltd., 1981): 17–42.

_____. "On Visiting the 'Moving Metropolis:' Reflections on the Architecture of Imperial Science," *Scientific Colonialism* (Washington, DC: Smithsonian Institution Press, 1987): 217–249.

_____. "Passages in Imperial Science: From Empire to Commonwealth." *Journal of World History* 4:1 (1993): 117–150.

Malin, S. R. C. and D. R. Barraclough. "Humboldt and the Earth's Magnetic Field." *Quarterly Journal of the Royal Astronomical Society* 32:3 (1991): 279–293.

Miller, David. "The Revival of the Physical Sciences in Britain: 1815–1840." *Osiris* 2 (1986): 107–134.

Millikan, Frank. "Joseph Henry's Grand Meteorological Crusade." *Weatherwise* 50 (October–November 1997): 14–17.

Morrell, Jack and Arnold Thackray. *Gentlemen of Science* (Oxford: Clarendon Press, 1981).

Moyer, Albert. *Joseph Henry* (Washington, DC: Smithsonian Institution Press, 1997).

Multhauf, Robert and Gregory Good. *A Brief History of Geomagnetism* (Washington, DC: Smithsonian Institution Press, 1987).

Musselman, Elizabeth Green. "Swords Into Ploughshares: John Herschel's Progressive View of Astronomical and Imperial Governance." *British Journal for the History of Science* 31:4 (December 1998): 419–436.

Pancaldi, Giuliano. "Scientific Internationalism and the British Association," *The Parliament of Science* (Northwood: Science Reviews Ltd., 1981): 145–169.

Philbrick, Nathaniel. *Sea of Glory* (New York: Viking, 2003).

Portolano, Marlana. "John Quincy Adams's Rhetorical Crusade for Astronomy." *Isis* 91:3 (September 2000): 480–503.

Pyenson, Lewis. *Civilizing Mission* (Baltimore: Johns Hopkins University Press, 1993).

_____. "Cultural Imperialism and Exact Sciences Revisited." *Isis* 84:1 (March 1993): 103–108.

_____. "Science and Imperialism," *Companion to the History of Modern Science* (London: Routledge, 1990): 920–931.

Reingold, Nathan. *Science American Style* (New Brunswick: Rutger's University Press, 1991).

Robinson, P. R. "Geomagnetic Observatories in the British Isles." *Vistas in Astronomy* 26 (1982): 347–367.

Ruskin, Steven. *John Herschel's Cape Voyage* (Burlington: Ashgate, 2004).

Savours, Ann and Anita McConnell. "The History of the Rossbank Observatory, Tasmania." *Annals of Science* 39:6 (November 1982): 527–564.

Schaffer, Simon. "Metrology, Metrication, and Victorian Values," *Victorian Science in Context* (Chicago: University of Chicago Press, 1997): 438–467.

Schroeder-Gudehus, Brigitte. "Nationalism and Internationalism," *Companion to the History of Modern Science* (London: Routledge, 1990): 909–919.

Shapin, Steven. *A Social History of Truth* (Chicago: University of Chicago, 1994).

Sibley, Joel. *The Partisan Imperative: The Dynamics of American Politics before the Civil War* (New York: Oxford University Press, 1985).

Smith, Geoffrey Sutton. "The Navy before Darwinism: Science, Exploration and Diplomacy in Antebellum America." *American Quarterly* 28:1 (Spring 1976): 41–55.

Snyder, Laura. "It's *All* Necessarily So: William Whewell on Scientific Truth." *Studies in History and Philosophy of Science* 25:5 (October 1994): 785–807.

Sobel, Dava. *Longitude* (New York: Walker, 1995).

Stafford, Robert. "Annexing the Landscapes of the Past: British Imperial Geology in the Nineteenth Century," *Imperialism and the Natural World* (Manchester: Manchester University Press, 1990): 67–89.

_____. *Scientist of Empire* (Cambridge: Cambridge University Press, 1989).

Stanton, William. *The Great United States Exploring Expedition* (Berkeley: University of California Press, 1975).

Theodorides, Jean. "Humboldt and England." *British Journal for the History of Science* 3:9 (1966): 39–55.

Warner, Deborah. "Terrestrial Magnetism: For the Glory of God and the Benefit of Mankind." *Osiris* 9 (1994): 67–84.

Webb, Robert K. *Modern England* (New York: Harper & Row, 1980).

Wettersten, John. "William Whewell: Problems of Induction vs. Problems of Rationality." *British Journal for the Philosophy of Science* 45:2 (June 1994): 716–742.

Williams, L. Pearce. *Michael Faraday* (New York: Basic Books, 1965).

Woolf, Harry. *The Transits of Venus* (Princeton: Princeton University Press, 1959).

Worboys, Michael. "The British Association and Empire: Science and Social Imperialism, 1880–1940," *The Parliament of Science* (Northwood: Science Reviews Ltd., 1981): 170–187.

Yeo, Richard. *Defining Science* (Cambridge: Cambridge University Press, 1993).

_____. "An Idol of the Market-Place: Baconianism in Nineteenth Century Britain." *History of Science* 23 (1985): 251–289.

_____. "Scientific Method and the Image of Science, 1831–1890," *The Parliament of Science* (Northwood: Science Reviews Ltd., 1981): 65–88.

_____. "Scientific Method and the Rhetoric of Science in Britain: 1830–1917," *Politics and Rhetoric of Scientific Method* (Dordrecht: Reidel Publishing, 1986): 259–289.

Zeller, Suzanne. "The Colonial World as Geological Metaphor: Strata(gems) of Empire in Victorian Canada." *Osiris* 15 (2001): 85–107.

Index

www.ingramcontent.com/pod-product-compliance
Lightning Source LLC
Chambersburg PA
CBHW061755260326
41914CB00006B/1113